# Lecture Notes in Mathematics 1624

Editors:
A. Dold, Heidelberg
F. Takens, Groningen

Subseries:
Mathematisches Institut der Universität
und Max-Planck-Institut für Mathematik
Bonn – vol. 22

Adviser:
F. Hirzebruch

**Springer**
*Berlin*
*Heidelberg*
*New York*
*Barcelona*
*Budapest*
*Hong Kong*
*London*
*Milan*
*Paris*
*Santa Clara*
*Singapore*
*Tokyo*

Vasile Brînzănescu

# Holomorphic Vector Bundles over Compact Complex Surfaces

 Springer

Author

Vasile Brînzănescu
Institute of Mathematics of Romanian Academy
P.O. Box 1-764
70700 Bucharest, Romania

Cataloging-in-Publication Data applied for

Die Deutsche Bibliothek - CIP-Einheitsaufnahme

**Brizănescu, Vasile:**
Holomorphic vector bundles over compact complex surfaces /
Vasile Brizănescu. - Berlin ; Heidelberg ; New York ;
Barcelona ; Budapest ; Hong Kong ; London ; Milan ; Paris ;
Tokyo : Springer, 1996
   (Lecture notes in mathematics ; 1624)
   ISBN 3-540-61018-9
NE: GT

Mathematics Subject Classification (1991):
32L10, 32J15, 32G13, 14D20, 14J05, 32L07, 53C07, 55R10

ISBN 3-540-61018-9 Springer-Verlag Berlin Heidelberg New York

© Springer-Verlag Berlin Heidelberg 1996
Printed in Germany

Typesetting: Camera-ready T$_E$X output by the author
SPIN: 10479984       46/3142-543210 - Printed on acid-free paper

To Ligia and Oana

# Introduction

The theory of holomorphic vector bundles over compact complex analytic manifolds is not completely developed. Still, in the case of compact complex surfaces, very important results were obtained in the last twenty years. The main purpose of this book is to present the available (sometimes only partial !) solutions to the two fundamental problems: the existence problem and the classification problem for holomorphic vector bundles over compact complex surfaces, with special emphasis to the case of nonalgebraic surfaces. Thus, the book is an introduction to the subject of holomorphic vector bundles over the nonalgebraic surfaces and the intersection with the classical books by C. Okonek, M. Schneider, H. Spindler *Vector bundles on complex projective spaces*, S. Kobayashi *Differential geometry of complex vector bundles*, R. Friedman, J.W. Morgan *Smooth four-manifolds and complex surfaces* is reasonably small.

The existence problem asks whether in a given ($\mathcal{C}^\infty$) topological vector bundle $E$ over a compact complex surface $X$ it exists a holomorphic structure. For an algebraic surface $X$, Schwarzenberger [Sw1] has shown that every topological vector bundle whose first Chern class lies in the Neron-Severi group of $X$, admits an algebraic structure. In the case of a nonalgebraic surface, the analogous condition $c_1(E) \in NS(X)$ (which is always necessary) is no longer sufficient, as the works of Elencwajg-Forster [E-F] and Bănică-Le Potier [B-L] show. We recall that a holomorphic vector bundle of rank $r$ over a complex manifold is irreducible if it admits no coherent analytic subsheaf of rank $k$, with $0 < k < r$. Irreducible vector bundles do not exist over algebraic manifolds, but there are plenty of them over compact complex nonalgebraic surfaces ( [E-F, B-L]). A complete answer to the existence problem in the nonalgebraic case is still unknown and the difficulty consists in the lack of a "good" technique to study irreducible vector bundles. The powerful method of Serre in the algebraic case produces (as expected) in the nonalgebraic case only the filtrable holomorphic structures in vector bundles.

Let $X$ be a compact complex surface and let $E$ be a $\mathcal{C}^\infty$ topological vector bundle over $X$. Let $P$ be a property (simple, stable, Hermite-Einstein, etc.) refering to holomorphic vector bundles. A family of $P$-holomorphic structures in $E$ parametrised by a complex space $S$ is a holomorphic vector bundle $\mathcal{E}$ over $S \times X$ such that:
(1) the restriction $\mathcal{E}_s$ of $\mathcal{E}$ to the fibre $X_s$ for every $s \in S$ is topologically (smoothly) isomorphic to $E$, and
(2) $\mathcal{E}_s$ has property $P$ for every point $s \in S$.
Two such families $\mathcal{E}, \mathcal{E}'$ are called equivalent if there exists a line bundle $\mathcal{L} \in Pic(S)$ such that $\mathcal{E}' \cong \mathcal{E} \otimes p^*(\mathcal{L})$, where $p : S \times X \to S$ is the first projection.

The classification problem asks if it is possible to endow the set of isomorphism classes of $P$-holomorphic structures in $E$ with a complex structure. More precisely, it asks if the functor which associates to every complex space $S$ the set of equivalence classes of families of $P$-holomorphic structures in $E$ parametrised by $S$, has a coarse moduli space or even a fine moduli space. It is possible to construct moduli spaces for simple vector bundles (see [L-O1, K-O, Nr2, Mj]), but these spaces are in general only locally Hausdorff.

In the case of an algebraic surface $X$ we have the notion of $H$-stability (Mumford-Takemoto) for vector bundles, where $H$ is a very ample line bundle over $X$. Maruyama [Mr3] and Gieseker [Gi1] proved that for stable vector bundles there exists a coarse moduli space. In the case of a Kähler surface $(X, \omega)$ we have the notion of $\omega$-stability and the $\omega$-stable vector bundles of a fixed topological type $E$ have moduli space which is a globally Hausdorff complex analytic space ( [L-O1]). More generally, on every compact complex surface there exist Gauduchon metrics $g$ and the notion of $g$-stability can be defined (an idea of Hitchin). The stability of Hermite-Einstein vector bundles was proved by Kobayashi [Ko2] and Lübke [Lu2]. The converse has been conjectured independently by Kobayashi and Hitchin. The Kobayashi-Hitchin correspondence relates the complex geometry concept of stable holomorphic vector bundle to the differential geometry concept of Hermite-Einstein connection. The existence of Hermite-Einstein connections in stable vector bundles was proved by Donaldson [Do2] for projective surfaces and later for projective manifolds, by Uhlenbeck and Yau [U-Y] for Kähler manifolds, by Buchdahl [Bh] for Gauduchon surfaces and finally, by Li and Yau [L-Y] for Gauduchon manifolds. For any $\mathcal{C}^\infty$ vector bundle $E$ over a compact complex manifold endowed with a Gauduchon metric $g$, there exists a coarse moduli space for $g$-stable holomorphic vector bundles with $\mathcal{C}^\infty$ support $E$ and this moduli space is an open Hausdorff subspace of the moduli space of simple vector bundles (see [Te1]).

In Chapter 1 we review some concepts and results on vector bundles and, more generally, on coherent analytic sheaves over compact complex manifolds. In Chapter 2 we give an outline of the Enriques-Kodaira classification of compact complex surfaces as well as examples in each class of surfaces with emphasis on the nonalgebraic case. Chapter 3 is devoted to the study of line bundles over compact complex surfaces. In Chapter 4 we present the available (partial) solutions to the existence problem for holomorphic vector bundles over compact complex surfaces. Chapter 5 is devoted to the classification problem for holomorphic vector bundles over compact complex surfaces. Since the subject is too large we give only the general lines of the constructions of the coarse moduli spaces. The table of contents is fairly self-explanatory and, for a more detailed description of their contents, the reader can consult the introductions of the chapters.

The book should be of interest not only to experts but also to graduate students and researchers in mathematics. Some parts of the book were the subject of a lecture course on holomorphic vector bundles for graduate students at S.I.S.S.A. Trieste.

## Acknowledgements

This work would not have been possible without the enlightening discussions on the subject of vector bundles with the late friend and colleague Constantin Bănică.

Parts of this book were elaborated during visits to the Ludwig-Maximilians-Universität in München, SFB Geometrie und Analysis in Göttingen, Max-Planck-Institut für Mathematik in Bonn and International Centre for Theoretical Physics in Trieste. The author would like to thank all these Institutions and Alexander von Humboldt-Stiftung for their support and for their kind hospitality. The author wants to express his warmest thanks to Professors O. Forster, H. Flenner, H. Grauert, F. Hirzebruch and M.S. Narasimhan for their help and for giving him the possibility of visiting these Institutions.

He is particularly grateful to Professor F. Hirzebruch for his interest and his help.

April 1995                    V. Brînzănescu

# Table of Contents

# 1. Vector bundles over complex manifolds

In this chapter we review, for the convenience of the reader, some concepts and results on vector bundles and, more generally, on coherent sheaves over (compact) complex manifolds. We present Chern classes, the Hirzebruch-Riemann-Roch Theorem, elementary properties of torsion-free and reflexive coherent sheaves and Serre GAGA type theorems. Finally, we put the fundamental problems on vector bundles over complex manifolds: the existence of holomorphic structures in a ($\mathcal{C}^\infty$) topological complex vector bundle over a complex manifold and the classification of holomorphic vector bundles over a complex manifold (the description of moduli spaces of holomorphic structures in a topological complex vector bundle).

## 1.1 Vector bundles

Let $X$ be a topological space. A *family of complex vector spaces* over $X$ is given by a topological space $E$ together with:

(1) a continuous map $p : E \to X$,

(2) a structure of finite dimensional complex vector space on every

$$E(x) = p^{-1}(x) \, , \, x \in X \, ,$$

compatible with the topology of $E(x)$, induced by the topology of $E$.

The topological space $E$ is called the *total space*, $X$ is called the *base space*, and $E(x)$ is called the *fibre* over $x \in X$.

A *morphism* of families of complex vector spaces over $X$

$$p : E \to X \, , \, p' : E' \to X$$

is given by a continuous map $\varphi : E \to E'$ such that

(1) $p' \circ \varphi = p$,

(2) for every $x \in X$ the restriction map $\varphi(x) : E(x) \to E'(x)$ is linear.

The morphism $\varphi : E \to E'$ is called an *isomorphism* if $\varphi$ is bijective and $\varphi^{-1} : E' \to E$ is continuous. Let $V$ be a finite dimensional complex vector space. The family $pr_1 : X \times V \to X$ is called the *product family* and any family of complex vector spaces isomorphic to a product family is called the *trivial family*.

Let $p : E \to X$ be a family of complex vector spaces over $X$ and let $Y \subset X$ be a topological subspace. Then $p : p^{-1}(Y) \to Y$ is a family of complex vector spaces over $Y$, called the *restriction* over $Y$ of the given family $E$ and denoted by $E|_Y$.

**Definition 1.1** A *(continuous) complex vector bundle* over $X$ is a family of complex vector spaces over $X$, $p : E \to X$, which is locally trivial, i.e. for every point $x \in X$ there exists an open neighbourhood $U$ $(x \in U \subset X)$ such that $E|_U$ is the trivial family.

A *morphism* of complex vector bundles over $X$ is a morphism of the corresponding families of complex vector spaces over $X$.

Let $p : E \to X$ be a complex vector bundle over $X$. Then, the dimension of the vector space $E(x)$, $x \in X$, is a locally constant function on $X$, hence constant if $X$ is connected. In this case, $r = \dim_{\mathbb{C}} E(x)$ is called the *rank* of the complex vector bundle $p : E \to X$. We shall suppose in the following that the topological space $X$ is connected.

Let $p : E \to X$ be a complex vector bundle of rank $r$ over $X$. It follows from the definition that there exist an open covering $(U_i)$ of $X$ and homeomorphisms

$$h_i : p^{-1}(U_i) \xrightarrow{\sim} U_i \times \mathbb{C}^r$$

over $U_i$ such that (for all $i$, $j$)

$$h_{ij} = h_i \circ h_j^{-1} : (U_i \cap U_j) \times \mathbb{C}^r \to (U_i \cap U_j) \times \mathbb{C}^r$$

is linear on each fibre. Thus,

$$h_{ij}(x, v) = (x, g_{ij}(x)v),$$

where

$$g_{ij} : U_i \cap U_j \to GL(r, \mathbb{C})$$

are continuous maps. The functions $(g_{ij})$ are called the *transition functions* of the vector bundle $p : E \to X$ (with respect to the *local trivializations* $(U_i, h_i)$).

On $U_i \cap U_j \cap U_k$ we have

$$g_{ij} g_{jk} g_{ki} = id. \tag{1.1}$$

Conversely, let $\mathcal{U} = (U_i)$ be an open covering of $X$ and let

$$g_{ij} : U_i \cap U_j \to GL(r, \mathbb{C})$$

be continuous maps with the property (1.1). Then, take the disjoint union $\coprod(U_i \times \mathbb{C}^r)$ and use $(g_{ij})$ to glue this to a complex vector bundle of rank $r$ over $X$

$$E := (\coprod(U_i \times \mathbb{C}^r))/ \sim$$

$$(x, v) \sim (x, g_{ij}(x)v).$$

If $X$ is a $(\mathcal{C}^\infty)$ differentiable manifold and all maps are differentiable one gets the notion of a *differentiable vector bundle*, using the natural differentiable structures on

$\mathbb{C}^r$, $GL(r, \mathbb{C})$.

If $X$ is an analytic space one defines a *holomorphic vector bundle* by requiring $E$ to be an analytic space and all maps to be holomorphic, using the natural holomorphic structures on $\mathbb{C}^r$, $GL(r, \mathbb{C})$.

To define *algebraic vector bundles* over algebraic varieties one requires everything to be algebraic , using the natural algebraic structure on $GL(r, \mathbb{C})$ as a Zariski open set of $\mathbb{C}^{r^2}$.

Let us work in the category of complex spaces and let us remark that everything works similarly in the other categories. Let $X$ be a (complex) connected, analytic space with structure sheaf $\mathcal{O}_X$. Denote by $\mathrm{Vect}^r_{hol}(X)$ the set of isomorphism classes of rank $r$ holomorphic vector bundles.

A *section* of the holomorphic vector bundle $p : E \to X$ over an open set $U \subset X$ is a holomorphic map $s : U \to E$ such that $p \circ s = id_U$.

Denote
$$\Gamma(U, E) := \{s \mid s : U \to E \text{ section}\}.$$
This defines a coherent analytic sheaf (of sections) $\mathcal{E}$ over $X$ by taking
$$\mathcal{E}(U) := \Gamma(U, E).$$
It is locally free, i.e.
$$\mathcal{E}|U_i \cong \mathcal{O}_{U_i}^{\oplus r}.$$
Conversely, let $\mathcal{E}$ be a locally free $\mathcal{O}_X$-module over $X$ and let
$$\varphi_i : \mathcal{E}|U_i \xrightarrow{\sim} \mathcal{O}_{U_i}^{\oplus r}$$
be local trivializations. We get the isomorphisms
$$\varphi_{ij} : \mathcal{O}_{U_i \cap U_j}^{\oplus r} \xrightarrow{\sim} \mathcal{O}_{U_i \cap U_j}^{\oplus r},$$
where $\varphi_{ij} = \varphi_i \circ \varphi_j^{-1}$, and the holomorphic maps
$$g_{ij} : U_i \cap U_j \to GL(r, \mathbb{C})$$
such that $\varphi_{ij,x}(u) = g_{ij}(x)u$, for every point $x \in U_i \cap U_j$. Obviously, one has $g_{ij}g_{jk}g_{ki} = id$ on $U_i \cap U_j \cap U_k$, i.e. $(g_{ij})$ is an $1 - cocycle$ of the covering $\mathcal{U} = (U_i)$ with coefficients in $GL(r, \mathcal{O}_X)$. Now, take the disjoint union $\coprod (U_i \times \mathbb{C}^r)$ and use $(g_{ij})$ to glue this to a vector bundle
$$E := (\coprod (U_i \times \mathbb{C}^r))/ \sim$$
$$(x, v) \sim (x, g_{ij}(x)v).$$

This proves the

**Proposition 1.2** *There is a bijection between* $\mathrm{Vect}^r_{hol}(X)$ *and the set of isomorphism classes of locally free sheaves of rank $r$ over $X$.*

*Remark.* In what follows we shall not distinguish between a vector bundle $E$ and the associated locally free sheaf $\mathcal{E}$ and, by abuse of notation, we shall use for both the same symbol $E$.

Let

$$p : E \to X \, , \, p' : E' \to X,$$

be holomorphic vector bundles over $X$, $E$ of rank $r$, $E'$ of rank $k$, and let $\varphi : E \to E'$ be a morphism of holomorphic vector bundles. Passing (if necessary) to a refinement of the coverings, we can suppose that there is a covering $\mathcal{U} = (U_i)$ of $X$ according to which both $E$ and $E'$ have trivializations and such that the map

$$U_i \times \mathbb{C}^r \to U_i \times \mathbb{C}^k$$

induced by the trivializations $h_i$ , $h'_i$ is of the form

$$(x, v) \longrightarrow (x, \varphi_i(x)v)$$

where

$$\varphi_i : U_i \longrightarrow \mathbb{C}^{rk} = \text{Mat}_{k \times r}(\mathbb{C}).$$

Note that the maps $\varphi_i$ must satisfy on $U_i \cap U_j$

$$\varphi_j g_{ij} = g'_{ij} \varphi_i.$$

Conversely, from a collection of maps $(\varphi_i)$ satisfying

$$\varphi_j g_{ij} = g'_{ij} \varphi_i$$

we can construct a morphism of vector bundles.

In case of the same rank $(k = r)$ we get from this that a necessary and sufficient condition for two 1-cocycles $(g_{ij})$ and $(g'_{ij})$ to define isomorphic vector bundles is the existence of maps $\varphi_i : U_i \to GL(r, \mathbb{C})$ such that

$$g'_{ij} = \varphi_j g_{ij} \varphi_i^{-1}.$$

This proves (see, for example, [Hr] ,p. 41) the

**Proposition 1.3** *There is a bijection between* $Vect_{hol}^r(X)$ *and the cohomology set* $H^1(X, GL(r, \mathcal{O}_X))$.

The above correspondence gives an easy way to define algebraic operations on vector bundles: direct sum $E \oplus F$, tensor product $E \otimes F$, dual bundle $E^*$, homomorphisms bundle $\mathcal{H}om(E, F)$, endomorphisms bundle $\mathcal{E}nd(E) := \mathcal{H}om(E, E)$, exterior power $\bigwedge^p E$, determinant bundle $\det(E) = \bigwedge^r E$ $(r = \text{rank}(E))$, symmetric power $S^p E$. For example, if the vector bundles $E$ , $F$ are represented by cocycles $(g_{ij})$ , $(g'_{ij})$, then the vector bundle $E \oplus F$ is represented by a cocycle $(h_{ij})$ where

$$h_{ij} = \begin{pmatrix} g_{ij} & 0 \\ 0 & g'_{ij} \end{pmatrix},$$

and the vector bundle $E \otimes F$ is represented by a cocycle $(t_{ij})$ where

$$t_{ij} = g_{ij} \otimes g'_{ij}$$

and where $\otimes$ denotes Kronecker product of matrices, etc.

Let $E$ be a holomorphic vector bundle over $X$ and $x \in X$ a point. We denote the stalk of $E$ (as coherent sheaf) at $x$ by $E_x$ and we set

$$E(x) = E_x \otimes_{\mathcal{O}_{X,x}} (\mathcal{O}_{X,x}/\boldsymbol{m}_x) = E_x/\boldsymbol{m}_x E_x \ ,$$

where $\boldsymbol{m}_x \subset \mathcal{O}_{X,x}$ denotes the maximal ideal of the local ring $\mathcal{O}_{X,x}$ (in fact $E(x)$ is the fibre of $E$ as vector bundle). A homomorphism of sheaves $\varphi : E \to F$ between two holomorphic vector bundles defines for every point $x \in X$ a homomorphism of $\mathcal{O}_{X,x}$-modules $\varphi_x : E_x \to F_x$ and a $\mathbb{C}$-linear map $\varphi(x) : E(x) \to F(x)$.

*Remark.* It can happen that $\varphi_x : E_x \to F_x$ is injective but $\varphi(x) : E(x) \to F(x)$ is not. If for example $X = \mathbb{C}$ and $z \in H^0(\mathbb{C}, \mathcal{O}_{\mathbb{C}})$ is the global section given by the coordinate function $z$ on $\mathbb{C}$, then

$$\varphi : \mathcal{O}_{\mathbb{C}} \to \mathcal{O}_{\mathbb{C}} \ , \ s \mapsto zs \ ,$$

is obviously a monomorphism of sheaves, but in the point $0 \in \mathbb{C}$

$$\varphi(0) : \mathcal{O}_{\mathbb{C},0}/\boldsymbol{m}_0 \to \mathcal{O}_{\mathbb{C},0}/\boldsymbol{m}_0$$

is the zero homomorphism.

On the other hand we have $\varphi_x : E_x \to F_x$ is surjective if and only if $\varphi(x) : E(x) \to F(x)$ is surjective, by Nakayama Lemma.

**Examples** We shall give some examples of vector bundles.
**(1)** Let $X$ be a complex space. Then, $pr_1 : X \times \mathbb{C}^r \to X$ is a holomorphic vector bundle. Any holomorphic vector bundle isomorphic to this one is called *holomorphic trivial vector bundle* of rank $r$.
**(2)** One of the more important is the tangent bundle on a complex manifold. Given a complex manifold $X$ we have a covering $(U_i)$ and holomorphic maps $\varphi_i : U_i \to \mathbb{C}^n$ which are biholomorphic isomorphisms onto open sets in $\mathbb{C}^n$. We define $g_{ij}$ on $U_i \cap U_j$ as the Jacobian matrix of

$$\varphi_i \circ \varphi_j^{-1} : \varphi_j(U_i \cap U_j) \to \varphi_i(U_i \cap U_j).$$

Clearly, these maps $(g_{ij})$ satisfy the compatibility conditions (1.1) for 1-cocycles, so a holomorphic vector bundle is defined on $X$, denoted by $T_X$ and called the *holomorphic tangent bundle*.
**(3)** On a complex manifold $X$ the coherent sheaf $\Omega^1_X$ is locally free and defines the holomorphic vector bundle of differentials or the *cotangent bundle*, since $\Omega^1_X$ is naturally isomorphic to the sheaf of sections of the dual of the holomorphic tangent bundle.
**(4)** Let $X$ be a complex manifold of dimension $n$. The *bundle of p-forms* is defined by $\Omega^p_X = \bigwedge^p \Omega^1_X$. For $p = n$ we get

$$\Omega^n_X = \bigwedge^n \Omega^1_X = \det(\Omega^1_X) = \omega_X (= \mathcal{K}_X),$$

the *canonical bundle* on $X$.

**(5)** Let $F : Y \to X$ be a morphism of complex manifolds and let $E$ be a holomorphic vector bundle over $X$ defined by a cocycle $(g_{ij})$ for a covering $\mathcal{U} = (U_i)$ of the manifold $X$. Then, obviously, the maps $(F^* g_{ij}) = (g_{ij} \circ F)$ for the covering $\mathcal{U}' = (F^{-1}(U_i))$ of $Y$ satisfy the compatibility conditions (1.1) for cocycles, so a holomorphic vector bundle is defined over $X$, denoted by $F^*E$ and called the *pull-back* of $E$ with respect to the map $F : Y \to X$.

**(6)** If $F : Y \to X$ is a morphism of complex manifolds then there is a natural morphism of holomorphic vector bundles over $Y$

$$\varphi : T_Y \to F^* T_X.$$

With local trivializations, we can consider $Y$ an open set in $\mathbb{C}^m$, $X$ an open set in $\mathbb{C}^n$ and $T_X$, $T_Y$, $F^* T_X$ trivial bundles. Then, the map $F$ is given by $n$ holomorphic functions $F_1, \cdots, F_n$ and the map $\varphi$ is given by the matrix $(\partial F_i / \partial z_j)$ ; $i = 1, \cdots, n$ ; $j = 1, \cdots, m$. Now, if $F : Y \to X$ is a closed imbedding of complex manifolds one sees easily that $T_Y$ is (locally) a direct summand of $F^* T_X$. Then there is a quotient bundle over $Y$,

$$N_{Y/X} := F^* T_X / T_Y,$$

called the *normal bundle* of $Y$ in $X$, of rank $\dim X - \dim Y = \mathrm{codim}_X Y$. If $\mathcal{I}$ is the sheaf of ideals defined by $Y$, then $\mathcal{I}/\mathcal{I}^2$ is a locally free $\mathcal{O}_Y$-module and we have an isomorphism of $\mathcal{O}_Y$-modules

$$\mathcal{H}om(\mathcal{I}/\mathcal{I}^2, \mathcal{O}_Y) \xrightarrow{\sim} N_{Y/X}.$$

The associated holomorphic vector bundle to $\mathcal{I}/\mathcal{I}^2$ over $Y$ is called, of course, the *conormal bundle* of $Y$ in $X$.

**(7)** An important class of vector bundles is the class of *line bundles*, i.e. vector bundles of rank one. According to the previous results the cohomology set classifying them is

$$H^1(X, GL(1, \mathcal{O}_X)) = H^1(X, \mathcal{O}_X^*).$$

Because the tensor product of two line bundles $L_1 \otimes L_2$ is a line bundle and since for the dual of a line bundle $L$, $L^* = \mathcal{H}om(L, \mathcal{O}_X)$, we have the isomorphism $L^* \otimes_{\mathcal{O}_X} L \xrightarrow{\sim} \mathcal{O}_X$, the set $H^1(X, \mathcal{O}_X^*)$ of isomorphism classes of line bundles has a structure of a group, called the *Picard group* of $X$ and denoted by $\mathrm{Pic}(X)$.

We shall give examples of holomorphic line bundles over complex projective spaces $\mathbb{P}^n$. The complex manifold $\mathbb{P}^n$ is the quotient of $\mathbb{C}^{n+1} \setminus \{0\}$ by the natural action of the multiplicative group $\mathbb{C}^*$

$$\mathbb{C}^{n+1} \setminus \{0\} \longrightarrow \mathbb{P}^n .$$

The *Hopf tautological line bundle* over $\mathbb{P}^n$ is the following subbundle of the trivial bundle of rank $(n+1)$ over $\mathbb{P}^n$:

$$\mathcal{O}_{\mathbb{P}^n}(-1) := \{(x, v) \in \mathbb{P}^n \times \mathbb{C}^{n+1} \mid v \in \ell_x , \; \ell_x \text{ the line defined by } x\}.$$

Taking the canonical covering $\mathcal{U} = (U_i)$ of $\mathbb{P}^n$, where $U_i = \{z_i \neq 0\}$, we get easily that the transition functions of the line bundle $\mathcal{O}_{\mathbb{P}^n}(-1)$ are $g_{ij} = z_i / z_j$ on $U_i \cap U_j$.

Let $(z_0 : z_1 : \cdots : z_n)$ be homogeneous coordinates on $\mathbb{P}^n$ and let $(z_0 : z_1 : \cdots : z_{n+1})$ be homogeneous coordinates on $\mathbb{P}^{n+1}$. Let $p = (0 : 0 : \cdots : 0 : 1)$ be a point in $\mathbb{P}^{n+1}$ and define the holomorphic (in fact, algebraic) map

$$\psi : \mathbb{P}^{n+1} \setminus \{p\} \longrightarrow \mathbb{P}^n , \ \psi(z_0 : \cdots : z_{n+1}) = (z_0 : \cdots : z_n)$$

(the projection of $\mathbb{P}^{n+1}$ from $p$ on $\mathbb{P}^n$ !). Clearly,

$$\psi^{-1}(z_0 : \cdots : z_n) = \{(z_0 : \cdots : z_n : \lambda) \mid \lambda \in \mathbb{C}\}$$

is a line. For $i = 0, 1, \cdots, n$ and $U_i = \{z_i \neq 0\} \subset \mathbb{P}^n$ we have

$$\psi^{-1}(U_i) \xrightarrow{\sim} U_i \times \mathbb{C}.$$

We get that $\psi : \mathbb{P}^{n+1} \setminus \{p\} \to \mathbb{P}^n$ is a line bundle over $\mathbb{P}^n$ and that the transition functions on $U_i \cap U_j$ are $g'_{ij} = z_j/z_i$. This line bundle is denoted by $\mathcal{O}_{\mathbb{P}^n}(1)$ and it is the dual bundle of the line bundle $\mathcal{O}_{\mathbb{P}^n}(-1)$ since $g'_{ij}g_{ij} = 1$.

For $k \in \mathbb{Z}$ we define the line bundles over $\mathbb{P}^n$

$$\mathcal{O}_{\mathbb{P}^n}(k) = \begin{cases} \mathcal{O}_{\mathbb{P}^n}(1)^{\otimes k} & \text{for } k \geq 0 \\ \mathcal{O}_{\mathbb{P}^n}(-1)^{\otimes(-k)} & \text{for } k < 0 \end{cases} .$$

*Remark.* From the definition of $\mathcal{O}_{\mathbb{P}^n}(k)$ it follows easily that its global sections are given by holomorphic functions $f : \mathbb{C}^{n+1} \setminus \{0\} \to \mathbb{C}$ with the property

$$f(\lambda z) = \lambda^k f(z) , \ \forall z \in \mathbb{C}^{n+1} \setminus \{0\} , \ \forall \lambda \in \mathbb{C}.$$

Since, by a theorem of Hartogs ($n + 1 \geq 2$ !) the function $f$ extends holomorphically on $\mathbb{C}^{n+1}$, it follows that $f = 0$ for $k < 0$. In particular, $\mathcal{O}_{\mathbb{P}^n}(-1)$ has no global sections and, the sections of $\mathcal{O}_{\mathbb{P}^n}(k)$ , $k \geq 0$ , can be identified with the homogeneous polynomials $P \in \mathbb{C}[z_0, \cdots, z_n]$ of degree $k$.

## 1.2  Chern classes

We shall define, following Hirzebruch [Hr], p. 58, the Chern (characteristic) classes $c_i(E) \in H^{2i}(X, \mathbb{Z})$ , $i = 0, 1, \cdots r$ , of a (continuous) complex vector bundle $E$ of rank $r$ over a paracompact topological space. The direct sum

$$H^*(X, \mathbb{Z}) = \oplus_i H^i(X, \mathbb{Z})$$

is a graded ring with respect to cup-product.

Let $\mathbb{P}^n$ be the complex projective space. The hyperplane $H = \{z_0 = 0\}$ in $\mathbb{P}^n$, with the induced orientation, is isomorphic to $\mathbb{P}^{n-1}$ and represents a $(2n - 2)$-dimensional integral homology class of $\mathbb{P}^n$, $[H] \in H_{2n-2}(\mathbb{P}^n, \mathbb{Z})$. Let $h \in H^2(\mathbb{P}^n, \mathbb{Z})$ be the corresponding cohomology class of $[H]$ by Poincaré duality

$$H_{2n-2}(\mathbb{P}^n, \mathbb{Z}) \cong H^2(\mathbb{P}^n, \mathbb{Z}).$$

The cohomology class $h$ is the "positive" generator of $H^2(\mathbb{P}^n, \mathbb{Z}) \cong \mathbb{Z}$.

We present the axioms for the Chern classes:

**Axiom 1** For every continuous complex vector bundle $E$ of rank $r$ over a paracompact space $X$ and every integer $i = 0, 1, \cdots, r$, there is a *Chern class* $c_i(E) \in H^{2i}(X, \mathbb{Z})$ and $c_0 = 1 \in H^0(X, \mathbb{Z}) \cong \mathbb{Z}$.
The element
$$c(E) = c_0(E) + c_1(E) + \cdots + c_r(E) \in H^*(X, \mathbb{Z})$$
is called the *total Chern class* of $E$.

**Axiom 2** (naturality) For any continuous map $f : Y \to X$ we have
$$c(f^*E) = f^*(c(E)),$$
where $f^* : H^*(X, \mathbb{Z}) \longrightarrow H^*(Y, \mathbb{Z})$ is the induced homomorphism.

**Axiom 3** If $E$ and $F$ are continuous complex vector bundles over $X$ then
$$c(E \oplus F) = c(E).c(F).$$

**Axiom 4** (normalisation) $c(\mathcal{O}_{\mathbb{P}^n}(1)) = 1 + h$ .

For the proof of the following theorem see Hirzebruch [Hr] ,p. 59:

**Theorem 1.4** *For every continuous complex vector bundle $E$ of rank $r$ over a paracompact space $X$ there exist and there are unique Chern classes $c_i(E) \in H^{2i}(X, \mathbb{Z})$ , $i = 0, 1, \cdots, r$ , which verify the Axioms 1-4.*

We shall present without details only two important steps.

*Step 1.* For every complex line bundle $L$ over a paracompact space $X$ there is a continuous map $f : X \to \mathbb{P}^n$ (for an appropriate $n$) such that $L \cong f^*(\mathcal{O}_{\mathbb{P}^n}(1))$. The idea is the following: for the continuous line bundle $L$ we can find global sections $\varphi_0, \cdots, \varphi_n \in H^0(X, L)$ such that they do not vanish simultaneously, and from these one gets a continuous map $f : X \to \mathbb{P}^n$ $(x \mapsto (\varphi_0(x) : \cdots : \varphi_n(x)))$. This map induces an isomorphism $L \cong f^*(\mathcal{O}_{\mathbb{P}^n}(1))$, where $\varphi_0, \cdots, \varphi_n$ are the pull-backs of the global sections $z_0, \cdots, z_n$ of the line bundle $\mathcal{O}_{\mathbb{P}^n}(1)$ on $\mathbb{P}^n$.

*Step 2.* (splitting principle) If $E$ is a continuous complex vector bundle of rank $r$ over a paracompact space $X$, there is another space $Y$ and a continuous map $f : Y \to X$ such that
(1)   $f^* : H^*(X, \mathbb{Z}) \longrightarrow H^*(Y, \mathbb{Z})$ is injective;
(2)   $f^*(E) \cong L_1 \oplus L_2 \oplus \cdots \oplus L_r$, $L_i$ line bundles for $i = 1, \cdots, r$.

We proceed by induction, the case $r = 1$ being trivial. Let $f_1 : \mathbb{P}(E) \to X$ be the associated projective bundle of lines in $E$ (i.e. a point of $\mathbb{P}(E)$ corresponds to a line through the origin of a fibre of $E$). This is constructed as follows: take the zero section of $E$

$$0 : X \to E \, , \ 0(x) = (x,0) \, , \ 0 \in E(x) \simeq \mathbb{C}^r$$

and consider on $E \setminus 0(X) = \bigcup_{x \in X}(E(x) \setminus \{0\})$ the action of $\mathbb{C}^*$ given by $\lambda(x,v) = (x, \lambda v)$. The quotient by this action is $\mathbb{P}(E)$ and $f_1$ is induced by the first projection. Any fibre of $f_1$ is isomorphic with $\mathbb{P}^{r-1}$. The vector bundle $f_1^*(E)$ over $\mathbb{P}(E)$ contains (as subbundle) a canonical line bundle $L \subset f_1^*(E)$

$$L := \{(\ell,v) \in \mathbb{P}(E) \times E \mid v \in \ell \, \}.$$

Then one gets an exact sequence of vector bundles over $\mathbb{P}(E)$

$$0 \to L \to f_1^*(E) \to f_1^*(E)/L \to 0.$$

An easy argument shows that this sequence splits (topologically), so $f_1^*(E) \cong L \oplus E_1$, where $E_1$ is the quotient vector bundle $f_1^*(E)/L$. By Leray-Hirsch theorem, see [Sp],Chapter 5, 7, we know that $f_1^* : H^*(X,\mathbb{Z}) \longrightarrow H^*(\mathbb{P}(E),\mathbb{Z})$ is injective. Now, by repeating this procedure we get the splitting principle.

*Remarks* (1) By using these two steps one can prove the existence and the uniqueness of the Chern classes.

(2) Differentiable and holomorphic vector bundles over appropriate spaces can be regarded as continuous vector bundles. Therefore Chern classes are defined in these cases too and they depend only on topological structure of these bundles.

We shall give another interpretation for the Chern class $c_1(L)$ of a continuous line bundle $L$ over a paracompact space $X$. Let $\mathcal{C}$ denote the sheaf of germs of local complex valued continuous functions on $X$ and let $\mathcal{C}^*$ denote the sheaf of germs of local non-zero complex valued continuous functions. The exponential homomorphism of sheaves $\mathcal{C} \overset{\exp}{\to} \mathcal{C}^*$ ($f \mapsto e^{2\pi i f}$) is surjective and the kernel is isomorphic to the constant sheaf $\mathbb{Z}$ over $X$. Therefore there is an exact sequence of sheaves

$$0 \to \mathbb{Z} \to \mathcal{C} \overset{\exp}{\to} \mathcal{C}^* \to 0.$$

Take the exact cohomology sequence

$$\cdots \to H^1(X,\mathcal{C}) \to H^1(X,\mathcal{C}^*) \overset{\delta}{\to} H^2(X,\mathbb{Z}) \to H^2(X,\mathcal{C}) \to \cdots .$$

Since the sheaf $\mathcal{C}$ is fine, the groups $H^1(X,\mathcal{C})$ and $H^2(X,\mathcal{C})$ are zero and we get the isomorphism:

$$H^1(X,\mathcal{C}^*) \cong H^2(X,\mathbb{Z}).$$

Therefore $\delta$ is an isomorphism between the group of (classes of) continuous line bundles over $X$ and the second integer cohomology group of $X$.

If $X$ is a differentiable manifold there exist an exact sequence

$$0 \to \mathbb{Z} \to \mathcal{C}_\infty \overset{\exp}{\to} \mathcal{C}_\infty^* \to 0,$$

where $\mathcal{C}_\infty$ is the sheaf of germs of local complex valued ($\mathcal{C}^\infty$) differentiable functions on $X$ (and correspondingly $\mathcal{C}_\infty^*$), and the exact cohomology sequence

$$\cdots \to H^1(X, \mathcal{C}_\infty) \to H^1(X, \mathcal{C}_\infty^*) \xrightarrow{\delta} H^2(X, \mathbb{Z}) \to H^2(X, \mathcal{C}_\infty) \to \cdots.$$

The sheaf $\mathcal{C}_\infty$ is fine and therefore $\delta$ is an isomorphism between the group of differentiable line bundles and $H^2(X, \mathbb{Z})$. It follows that the natural homomorphism

$$H^1(X, \mathcal{C}_\infty^*) \to H^1(X, \mathcal{C}^*)$$

is an isomorphism.

If $X$ is a complex (analytic) manifold there exists an exact sequence

$$0 \to \mathbb{Z} \to \mathcal{O}_X \xrightarrow{\exp} \mathcal{O}_X^* \to 0 \,,$$

which gives the long exact cohomology sequence:

$$\to H^1(X, \mathbb{Z}) \to H^1(X, \mathcal{O}_X) \to H^1(X, \mathcal{O}_X^*) \xrightarrow{\delta} H^2(X, \mathbb{Z}) \to H^2(X, \mathcal{O}_X) \to .$$

Let $\mathrm{Pic}_0(X) := \mathrm{Ker}(\delta)$; it follows that the group $\mathrm{Pic}(X)/\mathrm{Pic}_0(X)$ is isomorphic to a subgroup of the group $H^2(X, \mathbb{Z})$, called the *Neron-Severi group* of $X$ and denoted by $\mathrm{NS}(X)$. In this case it is no longer generally true that $\delta$ is an isomorphism.

We have the following interpretation for the Chern class $c_1(L)$ of a continuous line bundle $L$ over a paracompact space $X$.

**Proposition 1.5** *Let $L$ be a continuous line bundle over $X$. If $\delta : H^1(X, \mathcal{C}^*) \to H^2(X, \mathbb{Z})$ is the above isomorphism then $c_1(L) = \delta(\mathrm{cls}. L)$.*

Indeed, since $\delta$ commutes with maps it is sufficient to prove the result $\delta(\mathcal{O}_{\mathbb{P}^n}(1)) = h_n$, where $h_n = h$ is the positive generator of $H^2(\mathbb{P}^n, \mathbb{Z}) \cong \mathbb{Z}$. For $n \geq 2$ any linear embedding $j : \mathbb{P}^{n-1} \to \mathbb{P}^n$ induces an isomorphism

$$j^* : H^2(\mathbb{P}^n, \mathbb{Z}) \to H^2(\mathbb{P}^{n-1}, \mathbb{Z})$$

such that $j^* h_n = h_{n-1}$. Since $j^* \delta(\mathcal{O}_{\mathbb{P}^n}(1)) = \delta j^*(\mathcal{O}_{\mathbb{P}^n}(1))$ it is sufficient to prove that $\delta(\mathcal{O}_{\mathbb{P}^1}(1)) = h_1$ for the Riemann sphere $S^2 = \mathbb{P}^1$. This can be done with simplicial cohomology (see Hirzebruch [Hr], p. 62, for details).

If $X$ is a differentiable manifold and $E$ is a differentiable vector bundle of rank $r$ over $X$ we can construct *real Chern classes*

$$\tilde{c}(E) = 1 + \tilde{c}_1(E) + \cdots + \tilde{c}_r(E) \in H^*(X, \mathbb{R})$$

$\tilde{c}_i(E) \in H^{2i}(X, \mathbb{R})$, by means of differential-geometric methods (see [Gf, Ko3, Wl]). Let us give only the idea of this construction. Let $A$ be a connection in $E$ and $F_A$ its curvature. Locally, we denote the connection form and the curvature form of $A$ by $\omega$ and $\Omega$, respectively. Then $\Omega = d\omega + \omega \wedge \omega$ is a matrix of 2-forms and

$$\det(I_r - \frac{1}{2\pi i}\Omega) = 1 + \gamma_1(\Omega) + \cdots + \gamma_r(\Omega)$$

is a $GL(r, \mathbb{C})$-invariant symmetric polynomial. It follows that each $\gamma_k(\Omega)$ is a globally defined differential form of degree $2k$ with real coefficients. One proves that the cohomology class of $\gamma_k(\Omega)$ does not depend on the connection $A$. Defining

$$\tilde{c}_k(E) := [\gamma_k(\Omega)] \in H^{2k}(X, \mathbb{R}),$$

where $[\gamma_k(\Omega)]$ is the class in the de Rham cohomology, we get the *real total Chern class* of $E$

$$\tilde{c}(E) = 1 + \tilde{c}(E) + \cdots + \tilde{c}(E) \in H^*(X, \mathbb{R}).$$

One verifies similar axioms for these real Chern classes. If

$$j : H^*(X, \mathbb{Z}) \to H^*(X, \mathbb{R})$$

is the natural homomorphism and if $c(E) \in H^*(X, \mathbb{Z})$ is the (topological) total Chern class of $E$, then one can prove that $j(c(E)) \in H^*(X, \mathbb{R})$ verifies the axioms for real total Chern class. It follows that $\tilde{c}(E) = j(c(E))$.

Let $X$ be a paracompact space and let $E$ be a continuous vector bundle over $X$ with Chern classes $c_i \in H^{2i}(X, \mathbb{Z})$. By using the splitting principle one factors the total Chern class $c(E)$ "formally" in an appropriate ring extension of $H^*(X, \mathbb{Z})$ as

$$c(E) = \prod_{i=1}^{r}(1 + a_i).$$

**Definition 1.6** The *Chern character* of the vector bundle $E$ is the cohomology class

$$ch(E) := \sum_{i=1}^{r} e^{a_i} \in H^*(X, \mathbb{Z}) \otimes \mathbb{Q}.$$

**Definition 1.7** The *Todd class* of the vector bundle $E$ is the cohomology class

$$Td(E) := \prod_{i=1}^{r} \frac{a_i}{1 - e^{-a_i}} \in H^*(X, \mathbb{Z}) \otimes \mathbb{Q}.$$

Now let $X$ be an $n$-dimensional projective manifold over $\mathbb{C}$ and let $E$ be an algebraic vector bundle of rank $r$ over $X$. Denote

$$h^i(E) := \dim_{\mathbb{C}} H^i(X, E).$$

**Definition 1.8** The *Euler characteristic* of $E$ is

$$\chi(E) := \sum_{i=1}^{n}(-1)^i h^i(E).$$

**Definition 1.9** The *Chern classes* of $X$ are defined as

$$c_i(X) := c_i(T_X) \in H^{2i}(X, \mathbb{Z}).$$

We have the following fundamental result (see Hirzebruch [Hr], p. 155):

**Theorem 1.10 (Hirzebruch-Riemann-Roch)** *Let $X$ be an $n$-dimensional projective manifold over $\mathbb{C}$ and let $E$ be an algebraic vector bundle of rank $r$ over $X$. Then*
$$\chi(E) = [ch(E) . Td(T_X)]_n ,$$
*where $[ . ]_n :=$ the component in $H^{2n}(X, \mathbb{Z}) \cong \mathbb{Z}$.*

**Example** Let $X$ be a complex surface. Then

$$ch(E) = r + c_1(E) + \frac{1}{2}(c_1^2(E) - 2c_2(E)) ,$$

$$Td(E) = 1 + \frac{1}{2}c_1(E) + \frac{1}{12}(c_1^2(E) + c_2(E)) .$$

Applying Theorem 1.10 to $\mathcal{O}_X$ we find Noether's formula

$$\chi(\mathcal{O}_X) = \frac{1}{12}(c_1^2(X) + c_2(X)).$$

Then we can write

$$\chi(E) = \frac{1}{2}(c_1^2(E) - 2c_2(E)) + \frac{1}{2}c_1(E).c_1(X) + r\chi(\mathcal{O}_X).$$

Another fundamental result is the following (see [Sr2]):

**Theorem 1.11 (Serre duality)** *If $X$ is a compact, connected, complex manifold of dimension $n$ with canonical line bundle $\omega_X$, then we have for any holomorphic vector bundle $E$ over $X$*
$$H^q(X, E)^* \cong H^{n-q}(X, E^* \otimes \omega_X) .$$

Let $X$ be a compact, connected, complex manifold of dimension $n$. We shall extend the definition of Chern classes for any coherent analytic sheaf over $X$. Let $\mathcal{O}_X$ be the sheaf of holomorphic functions on $X$ and $\mathcal{A}$ the sheaf of $\mathbb{R}$-analytic complex valued functions. One knows that $\mathcal{A}$ is flat over $\mathcal{O}_X$ (see [Ma]). Let $\mathcal{F}$ be a coherent analytic sheaf over $X$; the $\mathcal{A}$-module $\mathcal{F} \otimes_{\mathcal{O}_X} \mathcal{A}$ has a finite resolution with free $\mathcal{A}$-modules $E_i$

$$0 \to E_{2n} \to \cdots \to E_0 \to \mathcal{F} \otimes_{\mathcal{O}_X} \mathcal{A} \to 0.$$

Then the *rank* of $\mathcal{F}$ is given by $\mathrm{rank}(\mathcal{F}) := \sum_i (-1)^i \mathrm{rank}(E_i)$ and the *total Chern class* is

$$c(\mathcal{F}) = 1 + c_1(\mathcal{F}) + \cdots + c_n(\mathcal{F}) := \prod_i c(E_i)^{(-1)^i} .$$

One can prove that $c(\mathcal{F})$ does not depend on the chosen resolution and, for an exact sequence of coherent analytic sheaves over $X$

$$0 \to \mathcal{F}' \to \mathcal{F} \to \mathcal{F}'' \to 0 ,$$

one has $c(\mathcal{F}) = c(\mathcal{F}').c(\mathcal{F}'')$. Also we can define the *Chern character* for a coherent analytic sheaf $\mathcal{F}$ over $X$:

$$ch(\mathcal{F}) = r + c_1(\mathcal{F}) + \frac{1}{2}(c_1^2(\mathcal{F}) - 2c_2(\mathcal{F})) + \cdots \in H^*(X, \mathbb{Q}).$$

*Remark.* Let $S$ be a complex analytic space and let $\mathcal{F}$ be a coherent analytic sheaf over $S \times X$, which is $S$-flat. For every point $s \in S$ one denotes by $\mathcal{F}_s$ the analytic sheaf induced over the fiber $X_s = \{s\} \times X$. Then, the rank and the Chern classes of the sheaf $\mathcal{F}_s$ are constant on the connected components of $S$. We say that the Chern classes are constant in flat families of sheaves.

By a result of Cartan-Serre (see [B-S], p. 110) we know that the cohomology groups of a compact complex manifold $X$ with values in a coherent analytic sheaf are finite dimensional $\mathbb{C}$-vector spaces, and we set

$$h^q(X, \mathcal{F}) := \dim_{\mathbb{C}} H^q(X, \mathcal{F}) .$$

The Hirzebruch-Riemann-Roch Theorem extends for coherent analytic sheaves over compact complex manifolds (see [B-T-T1])

$$\chi(\mathcal{F}) = [ch(\mathcal{F}).Td(T_X)]_n .$$

Let $f : X \to Y$ be a morphism of compact complex manifolds. One denotes by

$$Td(X/Y) := Td(T_X)/f^*(Td(T_Y))$$

the relative Todd class. If $\mathcal{F}$ is a coherent analytic sheaf over $X$, the direct image sheaves $R^i f_*(\mathcal{F})$ are coherent analytic sheaves (see [G-R3], [B-S], p. 99) over $Y$ and the element

$$f_!(\mathcal{F}) := \sum_i (-1)^i R^i f_*(\mathcal{F})$$

is defined in the Grothendieck group of $Y$. Its Chern character is given by the Grothendieck-Riemann-Roch formula (see [B-T-T2]):

$$ch(f_!(\mathcal{F})) = f_*(ch(\mathcal{F}).Td(X/Y)) .$$

If $\mathcal{F}$, $\mathcal{F}'$ are two coherent analytic sheaves over $X$ and if $\omega_X$ is the canonical line bundle over $X$ we have an extension of Serre duality: there exists a canonical pairing

$$\text{Ext}^i(\mathcal{F}, \mathcal{F}') \times \text{Ext}^{n-i}(\mathcal{F}', \mathcal{F} \otimes \omega_X) \to \mathbb{C} ,$$

which gives a duality between these two vector spaces (see [Ba]).

We shall use further the following

**Example** Let $X$ be a complex manifold of dimension $n$ and let $E$ be a holomorphic vector bundle of rank $r$ over $X$. Let $s \in \Gamma(X, E)$ be a holomorphic section and let $Z = \{s = 0\}$ be the set of zeros of $s$. If $Z$ has $\text{codim}_X Z = r$, we have

$$[Z] \in H_{2n-2r}(X, \mathbb{Z}) \cong H^{2r}(X, \mathbb{Z})$$

by Poincaré duality. Then $c_r(E) = [Z]$. In the case $n = r$, $Z$ is 0-dimensional and $c_r(E) = \text{length}(Z)$.

## 1.3 GAGA Theorems

Let $X$ be a complex projective algebraic variety. We can attach to $X$ an analytic space $X^{an}$ in a natural way such that there is a morphism

$$m : X^{an} \longrightarrow X$$

of ringed spaces. If $\mathcal{F}$ is any coherent sheaf on $X$ in the algebraic category, we can associate to $m^{-1}(\mathcal{F})$ a coherent sheaf $\mathcal{F}^{an}$ on $X^{an}$ in the analytic category by tensoring with $\mathcal{O}_{X^{an}}$. The stalk of this sheaf at a point $x \in X$ is

$$\mathcal{F}^{an}_x = \mathcal{O}^{an}_x \otimes_{\mathcal{O}_x} \mathcal{F}_x \, .$$

Thus $\mathcal{F}^{an} = m^*(\mathcal{F})$ in the category of ringed spaces. Algebraically, the important fact is that the natural homomorphism

$$\mathcal{O}_x \longrightarrow \mathcal{O}^{an}_x$$

is faithfully flat (both rings are noetherian local rings with the same completions); see [Mt], p. 13.

In general for a morphism of ringed spaces $f : X \to Y$ and a sheaf $\mathcal{F}$ on $Y$ there are induced functorial maps

$$H^i(Y, \mathcal{F}) \longrightarrow H^i(X, f^*(\mathcal{F})) \, .$$

In particular, for any $i$ there are natural maps

$$H^i(X, \mathcal{F}) \longrightarrow H^i(X^{an}, \mathcal{F}^{an}) \, .$$

Clearly, an $\mathcal{O}_X$-homomorphism between coherent algebraic sheaves $\mathcal{F} \to \mathcal{G}$ induces an $\mathcal{O}^{an}$-homomorphism $\mathcal{F}^{an} \to \mathcal{G}^{an}$. We get a natural map of sheaves

$$(\mathcal{H}om_{\mathcal{O}_X}(\mathcal{F}, \mathcal{G}))^{an} \longrightarrow \mathcal{H}om_{\mathcal{O}^{an}_X}(\mathcal{F}^{an}, \mathcal{G}^{an}) \tag{1.2}$$

which is an isomorphism. Indeed, at a point $x \in X$ this morphism is the natural map

$$\mathrm{Hom}_{\mathcal{O}_x}(\mathcal{F}_x, \mathcal{G}_x) \otimes_{\mathcal{O}_x} \mathcal{O}^{an}_x \to \mathrm{Hom}_{\mathcal{O}^{an}_x}(\mathcal{F}_x \otimes_{\mathcal{O}_x} \mathcal{O}^{an}_x, \mathcal{G}_x \otimes_{\mathcal{O}_x} \mathcal{O}^{an}_x) \, ,$$

which is an isomorphism by the faithfully flat property.

The comparison results between coherent algebraic sheaves $\mathcal{F}$ and coherent analytic sheaves $\mathcal{F}^{an}$ are called the *GAGA Theorems* (see [Sr3]).

**Theorem 1.12 (Serre)** *Let $X$ be a complex projective algebraic variety and $\tilde{\mathcal{F}}$ a coherent analytic sheaf on $X^{an}$. Then there is a unique coherent algebraic sheaf $\mathcal{F}$ on $X$ such that $\mathcal{F}^{an} \cong \tilde{\mathcal{F}}$. Moreover the natural maps*

$$H^i(X, \mathcal{F}) \longrightarrow H^i(X^{an}, \mathcal{F}^{an})$$

*are isomorphisms for all $i$.*

We shall give only a rough idea of the proof. By standard restrictions arguments ( [Sr3, Gf], p. 75) we get that it is sufficient to prove GAGA for $X = \mathbb{P}^n$. The key point is in the following two theorems:

**Theorem 1.13 (Theorem A of Cartan-Serre)** *If $\tilde{\mathcal{F}}$ is a coherent analytic sheaf on $(\mathbb{P}^n)^{an}$ there is an integer $d_0$ such that $\tilde{\mathcal{F}}(d) = \tilde{\mathcal{F}} \otimes_{\mathcal{O}_{\mathbb{P}^n}^{an}} \mathcal{O}_{\mathbb{P}^n}^{an}(d)$ is generated by its global sections for $d \geq d_0$ (i.e. for each $x \in \mathbb{P}^n$ there are $s_i \in H^0((\mathbb{P}^n)^{an}, \tilde{\mathcal{F}}(d))$ which generate $\tilde{\mathcal{F}}_x(d)$ as an $\mathcal{O}_{\mathbb{P}^n, x}^{an}$-module).*

**Theorem 1.14 (Theorem B of Cartan-Serre)** *If $\tilde{\mathcal{F}}$ is a coherent analytic sheaf on $(\mathbb{P}^n)^{an}$ there is an integer $d_0$ such that $H^i((\mathbb{P}^n)^{an}, \tilde{\mathcal{F}}(d)) = 0$ for $i > 0$, $d \geq d_0$.*

By using the computation of the cohomology groups $H^i((\mathbb{P}^n)^{an}, \mathcal{O}_{\mathbb{P}^n}^{an}(d))$ (see [Sr3, B-S], p. 139), induction on the dimension $n$ of the projective space (on $\mathbb{P}^0$ there is nothing to prove !) and descending induction on the order of the cohomology groups ($H^i((\mathbb{P}^n)^{an}, \tilde{\mathcal{F}}) = 0$ for $i > n$ !) one proves Theorems A and B by standard arguments with exact sequences (see, for details [Sr3, Gf], p. 79, [B-S], p. 147).

According to Theorem B there is, for any coherent analytic sheaf $\tilde{\mathcal{F}}$, an integer $d$ such that there is a surjective map

$$\oplus \mathcal{O}_{\mathbb{P}^n}^{an}(d) \to \tilde{\mathcal{F}} \to 0 .$$

Applying the same theorem to the kernel of this map we get $\tilde{\mathcal{F}}$ as a cokernel

$$\oplus \mathcal{O}_{\mathbb{P}^n}^{an}(d') \xrightarrow{\tilde{\varphi}} \oplus \mathcal{O}_{\mathbb{P}^n}^{an}(d) \to \tilde{\mathcal{F}} \to 0 .$$

Since $\mathcal{O}_{\mathbb{P}^n}^{an}(d) \cong (\mathcal{O}_{\mathbb{P}^n}(d))^{an}$ we get from the isomorphism (1.2) that the map $\tilde{\varphi}$ is induced by an algebraic map $\varphi$

$$\oplus \mathcal{O}_{\mathbb{P}^n}(d') \xrightarrow{\varphi} \oplus \mathcal{O}_{\mathbb{P}^n}(d) .$$

Then we take the coherent algebraic sheaf $\mathcal{F}$ to be the cokernel of $\varphi$

$$\oplus \mathcal{O}_{\mathbb{P}^n}(d') \xrightarrow{\varphi} \oplus \mathcal{O}_{\mathbb{P}^n}(d) \to \mathcal{F} \to 0 .$$

From the corresponding analytic exact sequence we get $\tilde{\mathcal{F}} \cong \mathcal{F}^{an}$. To show that a holomorphic sheaf is induced by a unique algebraic sheaf, suppose that $\mathcal{F}$, $\mathcal{G}$ are coherent algebraic sheaves such that $\mathcal{F}^{an} \cong \mathcal{G}^{an}$. By (1.2) there are

$$\varphi \in \mathrm{Hom}_{\mathcal{O}}(\mathcal{F}, \mathcal{G}) = H^0(\mathcal{H}om_{\mathcal{O}}(\mathcal{F}, \mathcal{G}))$$

and

$$\psi \in \mathrm{Hom}_{\mathcal{O}}(\mathcal{G}, \mathcal{F}) = H^0(\mathcal{H}om_{\mathcal{O}}(\mathcal{G}, \mathcal{F}))$$

inducing these isomorphisms, i.e. $\varphi^{an} \circ \psi^{an} = id$ and $\psi^{an} \circ \varphi^{an} = id$. Then, again from (1.2), it follows that $\varphi \circ \psi = id$, $\psi \circ \varphi = id$ algebraically.
By descending induction on $i$ one gets in a standard way the desired isomorphisms for cohomology.

**Corollary 1.15 (Chow Theorem)** *Every analytic subvariety of a projective variety is algebraic.*

If $X$ is an analytic subvariety of a projective algebraic variety then it is the support of a coherent analytic sheaf and thus (by GAGA) the support of a coherent algebraic sheaf.

The most important for our purpose is the following

**Corollary 1.16** *Every holomorphic vector bundle over a projective variety is induced by a unique algebraic vector bundle.*

It must be shown that if $\mathcal{F}$ is a coherent algebraic sheaf then $\mathcal{F}^{an}$ is locally free if and only if $\mathcal{F}$ is locally free. This follows from flatness.

## 1.4 Torsion-free and reflexive coherent sheaves

In this section we shall gather some useful results from the theory of coherent analytic sheaves on complex manifolds. For more details, see [O-S-S, Ko3]. This is motivated by the fact that torsion-free coherent analytic sheaves appear in a natural way in the study of holomorphic vector bundles over complex manifolds.

Let $X$ be a complex manifold of dimension $n$, $x \in X$ a point. The local noetherian ring $\mathcal{O}_{X,x}$, with maximal ideal $\boldsymbol{m}_x$, is regular ($\mathcal{O}_{X,x}$ is isomorphic to the ring of convergent power series in $n$ complex variables and $\boldsymbol{m}_x$ corresponds to the ideal of power series without constant term). Let $\mathcal{F}$ be a coherent sheaf over $X$. The stalk $\mathcal{F}_x$ is a finitely generated module over $\mathcal{O}_{X,x}$. The *homological dimension* of $\mathcal{F}_x$, denoted by $dh(\mathcal{F}_x)$, is defined to be the length of a minimal free resolution of $\mathcal{F}_x$. The *homological codimension* of $\mathcal{F}_x$, denoted by $codh(\mathcal{F}_x)$, is defined to be the maximal length of an $\mathcal{F}_x$- sequence in $\mathcal{O}_{X,x}$. For a proof of the following result see [Sr5], Chapter IV.

**Theorem 1.17 (syzygy)** $dh(\mathcal{F}_x) + codh(\mathcal{F}_x) = n.$

From the definition of homological dimension we know that $\mathcal{F}_x$ is a free $\mathcal{O}_{X,x}$-module if and only if $dh(\mathcal{F}_x) = 0$ or, equivalently. if and only if $codh(\mathcal{F}_x) = n$.

**Definition 1.18** For each integer $m$, $0 \leq m \leq n$ , the *m-th singularity set* of $\mathcal{F}$ is defined to be
$$S_m(\mathcal{F}) = \{x \in X \mid codh(\mathcal{F}_x) \leq m\}.$$

From the syzygy theorem it follows
$$S_m(\mathcal{F}) = \{x \in X \mid dh(\mathcal{F}_x) \geq n - m\}.$$

Obviously,

$$S_0(\mathcal{F}) \subset S_1(\mathcal{F}) \subset \cdots \subset S_{n-1}(\mathcal{F}) \subset S_n(\mathcal{F}) = X.$$

We call $S_{n-1}(\mathcal{F})$ the *singularity set* of $\mathcal{F}$ and we denote it by $S(\mathcal{F})$. It is clear that

$$S(\mathcal{F}) = \{x \in X \mid \mathcal{F}_x \text{ is not free over } \mathcal{O}_{X,x} \}.$$

For a proof of the following result, see [Sh]:

**Theorem 1.19** *The $m$-th singularity set $S_m(\mathcal{F})$ of a coherent sheaf is a closed analytic subset of $X$ of codimension $\geq n - m$.*

**Corollary 1.20** *The singularity set $S(\mathcal{F})$ of a coherent sheaf $\mathcal{F}$ is a closed analytic subset of $X$ of codimension at least 1.*

Over the open set $X \setminus S(\mathcal{F})$ the sheaf $\mathcal{F}$ is locally free. If $X$ is connected the *rank* of $\mathcal{F}$ is defined by

$$\mathrm{rk}(\mathcal{F}) = \mathrm{rank}(\mathcal{F}|_{X \setminus S(\mathcal{F})}).$$

Clearly, it coincides with the rank defined in Section 1.2.

**Definition 1.21** A coherent sheaf $\mathcal{F}$ over $X$ is a *$k$-th syzygy sheaf* if for every point $x$ of $X$ there is an open neighbourhood $U$ and an exact sequence

$$0 \rightarrow \mathcal{F}|_U \rightarrow E_1 \rightarrow \cdots \rightarrow E_k$$

such that $E_1, \cdots E_k$ are locally free coherent sheaves over $U$.

**Corollary 1.22** *If a coherent sheaf $\mathcal{F}$ is a $k$-th syzygy sheaf then*

$$\mathit{codim}\, S_m(\mathcal{F}) \geq n - m + k.$$

*In particular, for the singularity set $S(\mathcal{F})$ one has*

$$\mathit{codim}\, S(\mathcal{F}) \geq k + 1.$$

*Proof.* Let $U \subset X$ be an open neighbourhood (as in previous definition) of an arbitrary point $x \in X$. Let

$$\mathcal{F}_0 = \mathcal{F}|_U , \quad \mathcal{F}_i = \mathrm{Coker}(E_i \rightarrow E_{i+1}) , \quad i = 1, \cdots, k-1.$$

From the short exact sequences

$$0 \rightarrow \mathcal{F}_{i-1} \rightarrow E_i \rightarrow \mathcal{F}_i \rightarrow 0 , \quad i = 1, \cdots, k ,$$

we get

$$\mathrm{dh}(\mathcal{F}_{i,x}) = \begin{cases} 0 & \text{if } \mathcal{F}_{i,x} \text{ is free} \\ \mathrm{dh}(\mathcal{F}_{i-1,x}) + 1 & \text{otherwise.} \end{cases}$$

It follows that

$$S_m(\mathcal{F}_0) \subset S_{m-1}(\mathcal{F}_1) \subset S_{m-2}(\mathcal{F}_2) \subset \cdots \subset S_{m-k}(\mathcal{F}_k) .$$

Since codim $S_{m-k}(\mathcal{F}_k) \geq n - m + k$, by the above theorem, we obtain
codim $S_m(\mathcal{F}) \geq n - m + k$.

**Definition 1.23** A coherent sheaf $\mathcal{F}$ over $X$ is *torsion-free* if every stalk $\mathcal{F}_x$ is a torsion free $\mathcal{O}_{X,x}$-module (i.e. $as = 0$ for $a \in \mathcal{O}_{X,x}$, $s \in \mathcal{F}_x$, implies $a = 0$ or $s = 0$).

Every locally free sheaf is obviously torsion-free. Any coherent subsheaf of a torsion-free sheaf is again torsion-free. Conversely, we have

**Proposition 1.24** *If $\mathcal{F}$ is a torsion-free coherent sheaf of rank $r$, then it is locally a subsheaf of a free sheaf of rank $r$ (in particular, every torsion-free coherent sheaf is a 1-st syzygy sheaf).*

*Proof.* For each $x \in X$ let $K$ be the quotient field of $\mathcal{O}_{X,x}$. Since $\mathcal{F}_x$ is torsion-free, the natural map
$$i : \mathcal{F}_x \longrightarrow \mathcal{F}_x \otimes_{\mathcal{O}_{X,x}} K \cong K^r$$
is injective. Since $\mathcal{F}_x$ is of finite type, the injection $i$ followed by multiplication by some non-zero element $a \in \mathcal{O}_{X,x}$ (choose a suitable element $a$ to clear the "denominators") gives an injection
$$j : \mathcal{F}_x \longrightarrow \mathcal{O}_{X,x}^r .$$
Because $\mathcal{F}$ is coherent, $j$ extends to a homomorphism $j : \mathcal{F}|_U \to \mathcal{O}_U^r$ for a neighbourhood $U$ of $x$. Then $j$ is injective in a possible smaller neighbourhood of $x$.

Thus the torsion-free sheaves are precisely the 1-st syzygy sheaves. From the above results we obtain

**Corollary 1.25** *The singularity set of a torsion-free coherent sheaf is at least 2-codimensional. In particular, every torsion-free coherent sheaf over a curve (Riemann surface) is locally free (i.e. a holomorphic vector bundle).*

The *dual* of a coherent sheaf is defined to be the coherent sheaf
$$\mathcal{F}^* := \mathcal{Hom}_{\mathcal{O}_X}(\mathcal{F}, \mathcal{O}_X) .$$
There is a natural homomorphism of $\mathcal{F}$ into its double dual $\mathcal{F}^{**}$:
$$\mu : \mathcal{F} \longrightarrow \mathcal{F}^{**} .$$
Since $\mathcal{F}_x^* = \mathrm{Hom}_{\mathcal{O}_x}(\mathcal{F}_x, \mathcal{O}_x)$ for every $x \in X$, we have for the homomorphism
$$\mu_x : \mathcal{F}_x \longrightarrow \mathcal{F}_x^{**} ,$$
that for each $s \in \mathcal{F}_x$, $\mu_x(s) \in \mathrm{Hom}_{\mathcal{O}_x}(\mathcal{F}_x^*, \mathcal{O}_x)$ is given by
$$(\mu_x(s))(f) = f(s) , \ f \in \mathcal{F}_x^* .$$
Then
$$\mathrm{Ker} \, \mu_x = \bigcap_{f \in \mathcal{F}_x^*} \mathrm{Ker} \, f .$$

**Proposition 1.26** *For every point $x \in X$ and any coherent sheaf $\mathcal{F}$ on $X$ let $T(\mathcal{F}_x)$ be its torsion $\mathcal{O}_{X,x}$-submodule (i.e. the set of torsion elements of $\mathcal{F}_x$). Then*

$$T(\mathcal{F}_x) \;=\; \operatorname{Ker} \mu_x .$$

*Proof.* [(cf. [G-R1], p. 233, [Ko3], p. 150)] If $s \in T(\mathcal{F}_x)$, then $as = 0$ for some non-zero element $a \in \mathcal{O}_{X,x}$, and

$$af(s) = f(as) = f(0) = 0 \text{ for all } f \in \mathcal{F}_x^* .$$

Hence $f(s) = 0$ for all $f \in \mathcal{F}_x^*$ ($\mathcal{O}_{X,x}$ has no zero-divisors !). It follows $s \in \operatorname{Ker} \mu_x$. Conversely, let $s$ be an element of $\mathcal{F}_x$ not belonging to $T(\mathcal{F}_x)$. We must prove that $f(s) \neq 0$ for some $f \in \mathcal{F}_x^*$. Since $\mathcal{F}_x/T(\mathcal{F}_x)$ is, obviously, torsion-free, it is a submodule of a free module $\mathcal{O}_{X,x}^r$ (by the proof of Proposition 1.24). Thus, there is an injection

$$j : \mathcal{F}_x/T(\mathcal{F}_x) \longrightarrow \mathcal{O}_{X,x}^r .$$

Let $p : \mathcal{F}_x \to \mathcal{F}_x/T(\mathcal{F}_x)$ be the natural projection. Since $p(s) \neq 0$, $(j \circ p)(s)$ is a non-zero element of $\mathcal{O}_{X,x}^r$. Taking $q : \mathcal{O}_{X,x}^r \to \mathcal{O}_{X,x}$ a suitable projection, $f = q \circ j \circ p \in \mathcal{F}_x^*$ has the desired property.

The coherent subsheaf $\operatorname{Ker} \mu$ of $\mathcal{F}$, denoted by $\mathcal{T}(\mathcal{F})$, is called the *torsion subsheaf* of $\mathcal{F}$. It is clear that $\mu$ is injective if and only if $\mathcal{F}$ is torsion-free (i.e. $\mathcal{T}(\mathcal{F}) = 0$). A coherent sheaf $\mathcal{F}$ is a *torsion sheaf* if $\mathcal{T}(\mathcal{F}) \cong \mathcal{F}$.

**Definition 1.27** A coherent sheaf $\mathcal{F}$ is said to be *reflexive* if the homomorphism $\mu : \mathcal{F} \longrightarrow \mathcal{F}^{**}$ is an isomorphism.

Every locally free sheaf is obviously reflexive. Every reflexive sheaf is torsion-free.

**Proposition 1.28** *The singularity set of a reflexive coherent sheaf is at least 3-codimensional. In particular, every reflexive coherent sheaf over a complex surface is locally free (i.e. a holomorphic vector bundle).*

*Proof.* By Corollary 1.22 it suffices to show that reflexive sheaves are 2-nd syzygy sheaves. Take a presentation

$$\mathcal{O}_U^q \to \mathcal{O}_U^p \to \mathcal{F}^*|_U \to 0$$

over an open set $U \subset X$. Dualizing it, we obtain an exact sequence

$$0 \to \mathcal{F}^{**}|_U \to \mathcal{O}_U^p \to \mathcal{O}_U^q$$

with $\mathcal{F}|_U \cong \mathcal{F}^{**}|_U$, i.e. $\mathcal{F}$ is a 2-nd syzygy sheaf.

**Definition 1.29** A coherent sheaf $\mathcal{F}$ over $X$ is *normal* if for every open set $U$ in $X$ and every analytic set $A \subset U$ of codimension at least 2, the restriction map

$$\Gamma(U, \mathcal{F}) \longrightarrow \Gamma(U \setminus A, \mathcal{F})$$

is an isomorphism.

By second Riemann extension Theorem the structure sheaf is normal. We note that, by Proposition 1.24, the above restriction map is injective if $\mathcal{F}$ is torsion-free.

**Proposition 1.30** *A coherent sheaf is reflexive if and only if it is torsion-free and normal.*

*Proof.* If $\mathcal{F}$ is reflexive, then $\mathcal{F} \cong \mathcal{F}^{**} = \mathcal{H}om_{\mathcal{O}_X}(\mathcal{F}^*, \mathcal{O}_X)$. Since $\mathcal{O}_X$ is torsion-free and normal, the dual of any coherent sheaf is torsion-free and normal. It follows $\mathcal{F}$ torsion-free and normal.

Conversely, assume that $\mathcal{F}$ is torsion-free and normal. Since $\mathcal{F}$ is torsion-free, the natural map $\mu : \mathcal{F} \to \mathcal{F}^{**}$ is injective and the singularity set $A = S(\mathcal{F})$ is of codimension at least 2 by Corollary 1.25. For every open set $U$ in $X$, we have the following commutative diagram:

$$
\begin{array}{ccc}
\Gamma(U, \mathcal{F}) & \xrightarrow{\ \mu\ } & \Gamma(U, \mathcal{F}^{**}) \\
{\scriptstyle res} \downarrow & & \downarrow {\scriptstyle res} \\
\Gamma(U \setminus (U \cap A), \mathcal{F}) & \xrightarrow{\ \mu'\ } & \Gamma(U \setminus (U \cap A), \mathcal{F}^{**})
\end{array}
$$

The vertical arrows are isomorphisms since $\mathcal{F}$ is normal and since $\mathcal{F}^{**}$ is normal (as a dual). The horizontal map $\mu'$ is an isomorphism since $\mu : \mathcal{F} \to \mathcal{F}^{**}$ is an isomorphism outside the singularity set $A = S(\mathcal{F})$. Hence, the map

$$\mu : \Gamma(U, \mathcal{F}) \longrightarrow \Gamma(U, \mathcal{F}^{**})$$

is also an isomorphism, i.e. $\mathcal{F}$ is reflexive.

**Corollary 1.31** *The dual $\mathcal{F}^*$ of any coherent sheaf $\mathcal{F}$ is reflexive.*

**Proposition 1.32** *Let*

$$0 \to \mathcal{F}' \to \mathcal{F} \to \mathcal{F}'' \to 0$$

*be an exact sequence of coherent sheaves, where $\mathcal{F}$ is reflexive and $\mathcal{F}''$ is torsion-free. Then $\mathcal{F}'$ is normal and hence reflexive.*

*Proof.* Let $U \subset X$ be an open set and $A \subset U$ an analytic subset of codimension at least 2. Since $\mathcal{F}'$, $\mathcal{F}''$ are torsion-free, the restriction mappings

$$\Gamma(U, \mathcal{F}') \to \Gamma(U \setminus A, \mathcal{F}') \quad , \quad \Gamma(U, \mathcal{F}'') \to \Gamma(U \setminus A, \mathcal{F}'')$$

are injective. From the diagram

$$\begin{array}{ccccccc}
0 & \longrightarrow & \Gamma(U,\mathcal{F}') & \longrightarrow & \Gamma(U,\mathcal{F}) & \longrightarrow & \Gamma(U,\mathcal{F}'') \\
& & \downarrow & & \wr\downarrow & & \downarrow \\
0 & \longrightarrow & \Gamma(U \setminus A,\mathcal{F}') & \longrightarrow & \Gamma(U \setminus A,\mathcal{F}) & \longrightarrow & \Gamma(U \setminus A,\mathcal{F}'')
\end{array}$$

it follows by standard arguments that the restriction map

$$\Gamma(U,\mathcal{F}') \longrightarrow \Gamma(U \setminus A,\mathcal{F}')$$

is an isomorphism.

**Proposition 1.33** *A reflexive sheaf of rank 1 is a line bundle.*

*Proof.* (cf. [O-S-S], Lemma 1.1.15) By Propositions 1.24 and 1.30 it suffices to show that every normal sheaf of ideals $0 \neq J \subset \mathcal{O}_X$ is invertible. Let $x \in X$ and let

$$J_x = \boldsymbol{q}_1 \cap \cdots \cap \boldsymbol{q}_m$$

be an (irreducible) primary decomposition of $J_x$ in $\mathcal{O}_{X,x}$. Let $U$ be a Stein neighbourhood of $x$ such that there exist sheaves of ideals $Q_1, \cdots, Q_m \subset \mathcal{O}_U$ with

$$Q_{i,x} = \boldsymbol{q}_i \quad \text{for } i = 1, \cdots, m$$

and

$$J|_U = Q_1 \cap \cdots \cap Q_m .$$

We claim that

$$\operatorname{codim}_x(\operatorname{supp}(\mathcal{O}_U/Q_i), U) = 1 \quad \text{for } i = 1, \cdots, m .$$

Let us assume, for example, that $A = \operatorname{supp}(\mathcal{O}_U/Q_1)$ has codimension at least 2 in $U$. It follows that

$$\Gamma(U \setminus A, Q_1) = \Gamma(U \setminus A, \mathcal{O}) ,$$

Since $J$ is normal we get the isomorphism

$$\Gamma(U, J) \overset{\sim}{\to} \Gamma(U \setminus A, J) .$$

Thus, we obtain a commutative diagram

$$\begin{array}{ccc}
\Gamma(U,J) = \Gamma(U,Q_1) \cap \cdots \cap \Gamma(U,Q_m) & \hookrightarrow & \Gamma(U,Q_2) \cap \cdots \cap \Gamma(U,Q_m) \\
\wr\downarrow & & \uparrow \\
\Gamma(U \setminus A, Q_1) \cap \cdots \cap \Gamma(U \setminus A, Q_m) & \overset{\sim}{\to} & \Gamma(U \setminus A, Q_2) \cap \cdots \cap \Gamma(U \setminus A, Q_m)
\end{array}$$

It follows

$$\Gamma(U, Q_1) \cap \cdots \cap \Gamma(U, Q_m) = \Gamma(U, Q_2) \cap \cdots \cap \Gamma(U, Q_m) \,.$$

Since $U$ is Stein, we obtain that

$$\boldsymbol{q}_1 \cap \cdots \cap \boldsymbol{q}_m = \boldsymbol{q}_2 \cap \cdots \cap \boldsymbol{q}_m \,,$$

in contradiction to the irreducibility of the decomposition of $J_x$. Thus, for the prime ideals $\boldsymbol{p}_i = \mathrm{rad}(\boldsymbol{q}_i) \subset \mathcal{O}_{X,x}$, we have

$$\dim(\mathcal{O}_{X,x}/\boldsymbol{p}_i) = \dim X - 1$$

and therefore $\boldsymbol{p}_i = (f_i)$ is a principal ideal ($\mathcal{O}_{X,x}$ is factorial). It follows that $\boldsymbol{q}_i = (f_i^{k_i})$ for a suitable integer $k_i \geq 1$ and

$$J_x = \boldsymbol{q}_1 \cap \cdots \cap \boldsymbol{q}_m = (f_1^{k_1} \cdots f_m^{k_m})$$

is a principal ideal.

**Definition 1.34** Let $\mathcal{F}$ be a torsion-free sheaf of rank $r$ over $X$. The sheaf

$$\det(\mathcal{F}) := (\overset{r}{\bigwedge} \mathcal{F})^{**}$$

is called the *determinant bundle* of $\mathcal{F}$.

By the above proposition $\det(\mathcal{F})$ is in fact a line bundle over $X$.

**Proposition 1.35** *A monomorphism $\mathcal{F} \hookrightarrow \mathcal{F}'$ between torsion-free sheaves of the same rank induces a monomorphism $\det(\mathcal{F}) \hookrightarrow \det(\mathcal{F}')$ of the determinant bundles.*

*Proof.* Outside of the analytic subset $A = S(\mathcal{F}') \cup S(\mathcal{F}'/\mathcal{F})$ the map $\mathcal{F} \to \mathcal{F}'$ is an isomorphism and thus also $\det(\mathcal{F}) \to \det(\mathcal{F}')$ is an isomorphism. Therefore $\mathrm{Ker}(\det(\mathcal{F}) \to \det(\mathcal{F}'))$ is a torsion sheaf over $X$ and, as a subsheaf of a torsion-free sheaf, it must be zero.

The definition of the determinant line bundle can be extended to any coherent sheaf $\mathcal{F}$ by using locally free resolutions of $\mathcal{F}|_U$, where $U \subset X$ is a small open set (see, for example [Ko3], p. 162). Then, we have ( [Ko3], p. 165, 166):

**Proposition 1.36** *If*
$$0 \to \mathcal{F}' \to \mathcal{F} \to \mathcal{F}'' \to 0$$
*is an exact sequence of coherent sheaves, then there is a canonical isomorphism*

$$\det(\mathcal{F}) \cong \det(\mathcal{F}') \otimes \det(\mathcal{F}'') \,.$$

**Proposition 1.37** *If $\mathcal{F}$ is a torsion sheaf, then $\det(\mathcal{F})$ admits non-trivial holomorphic sections. Moreover, if*

$$\mathrm{supp}(\mathcal{F}) = \{x \in X \mid \mathcal{F}_x \neq 0\,\}$$

*has codimension at least 2, then det($\mathcal{F}$) is trivial.*

Let $\mathcal{F} \subset \mathcal{E}$ be a coherent subsheaf of the reflexive sheaf $\mathcal{E}$. Take the quotient $Q = \mathcal{E}/\mathcal{F}$ and let $q : \mathcal{E} \to Q$ be the natural map. Let $\hat{\mathcal{F}} := q^{-1}(\mathcal{T}(Q))$, where $\mathcal{T}(Q)$ is the torsion subsheaf of $Q$. Obviously, $\mathcal{F} \subset \hat{\mathcal{F}} \subset \mathcal{E}$, $\mathrm{rk}(\hat{\mathcal{F}}) = \mathrm{rk}(\mathcal{F})$ and the quotient $\mathcal{E}/\hat{\mathcal{F}}$ is torsion-free. By Proposition 1.32 we get that $\hat{\mathcal{F}}$ is reflexive.

**Definition 1.38** The sheaf $\hat{\mathcal{F}}$ is called the *maximal reflexive extension* of $\mathcal{F}$ in the reflexive sheaf $\mathcal{E}$.

## 1.5 Problems on vector bundles

Let $X$ be a compact complex manifold. We shall present the fundamental problems for holomorphic vector bundles over $X$. Firstly, the *existence problem* of holomorphic structures in topological (smooth) vector bundles.

**Problem I (existence)** *Which complex topological vector bundles over $X$ do admit holomorphic structures ?*

For line bundles (rank one case) the problem is solved by the exponential exact sequence. From Section 1.2 we know that $H^2(X, \mathbb{Z})$ ($\cong H^2(X, \mathcal{C}^*)$) parametrises the topological (or smooth) line bundles, and thus the natural homomorphism $H^1(X, \mathcal{O}_X^*) \overset{\delta}{\to} H^2(X, \mathbb{Z})$ maps the holomorphic structure into the induced topological structure. Therefore, a topological line bundle $L$ has a holomorphic structure if and only if $\delta(L) = c_1(L)$ belongs to the Neron-Severi group $\mathrm{NS}(X)$.

Let us suppose that the rank $r \geq 2$. In order to solve the existence problem we have to give firstly the topological classification of vector bundles of rank $r$. The Chern classes represent a first approximation of the topological type of a vector bundle. If the vector bundle $E$ is topologically trivial, then $c_i(E) = 0$ for all $i \geq 1$. The converse is not generally true. We mention some results for the case $X$ is a finite CW-complex of (real) dimension $n$:

(a) For a natural number $r$ one denotes by $\mathrm{Vect}_{top}^r(X)$ the isomorphism classes of complex topological vector bundles over $X$, of rank $r$. Then the map

$$\mathrm{Vect}_{top}^{[n/2]}(X) \longrightarrow \mathrm{Vect}_{top}^r(X) \quad , \quad E \to E \oplus (r - [n/2])\mathbb{1} \; ,$$

is onto whenever $r \geq [n/2]$ (where $[n/2]$ is the integer part of $n/2$ and $(r-[n/2])\mathbb{1}$ is the trivial vector bundle of rank $r - [n/2]$). Moreover, for $r \geq n/2$ the same map is bijective; see for instance [Hs], p. 99.

(b) If $r \geq n/2$ and $H^{2q}(X, \mathbb{Z})$ has not $(q-1)!$-torsion for any $q \geq 1$, then two vector bundles $E$ and $E'$ of rank $r$ over $X$ are isomorphic if and only if they have the same Chern classes (see [Pe]).

Let us suppose now that $X$ is a compact Riemann surface (complex curve). From the above results we get in this case the topological classification: the isomorphism classes of complex topological vector bundles are parametrised by the set

$\{(r, c_1) \mid r$ a positive integer , $c_1$ an integer $\}$, where $r$ is the rank and $c_1$ is the first Chern class (or the degree) of the complex vector bundle. Any topological vector bundle $E$ of rank $r \geq 2$ has the form $E \cong L \oplus (r-1)\mathbb{1}$, where $L$ is a line bundle and $(r-1)\mathbb{1}$ is the trivial topological vector bundle of rank $r-1$. Since $H^2(X, \mathcal{O}_X) = 0$ in this case, we get from the exact cohomology sequence associated to the exponential sequence (see Section 1.2) that every topological line bundle has a holomorphic structure. It follows that every topological vector bundle over a curve has a holomorphic structure.

If $X$ is a surface (compact, connected complex manifold of dimension two) the topological classification of complex vector bundles was given by Wu (see [Wu]): the isomorphism classes of complex topological vector bundles are parametrised by the set

$$\{(r, c_1, c_2) \mid r \text{ an integer } \geq 2, \ c_1 \in H^2(X, \mathbb{Z}) \ , \ c_2 \in H^4(X, \mathbb{Z}) \ \}.$$

We shall give in Chapter 4 the present state of knowledge on the existence problem for holomorphic structures in a given topological (smooth) vector bundle over a surface and we shall explain the difficulty of this problem. Here we only mention that a complete answer to the existence problem in the nonalgebraic (surfaces) context is still unknown.

*Remark.* If $X$ is a complex projective manifold then, by GAGA (see Section 1.3), the existence of a holomorphic structure in a given topological vector bundle is equivalent to the existence of an algebraic structure.

Let us consider the case dim $X = 3$. If $E$ is topological vector bundle of rank $r \geq 3$, then the Chern classes $(c_1(E), c_2(E), c_3(E))$ determine up to an isomorphism the bundle $E$. This fact follows by the above mentioned results and it was firstly proved in [Bc]. If $r = \mathrm{rk}(E) > 3$, then $E \cong E' \oplus (r-3)\mathbb{1}$, where $E'$ is a vector bundle of rank 3, uniquely determined (up to an isomorphism) by the isomorphism class of $E$. Therefore, the topological classification is interesting for the rank 2 and 3 and it was given by Atiyah-Rees [A-R] and Bănică-Putinar [B-P]. As to the existence problem we mention only the following result: "Every topological vector bundle over $\mathbb{P}^3$ has a holomorphic structure". It is due to Horrocks [Ho], Atiyah-Rees [A-R] and Vogelaar [Vg]. For more results on this problem see Bănică-Putinar [B-P], Bănică-Coandă [B-C] and Okonek-Schneider-Spindler [O-S-S].

In the case dim $X \geq 4$ the situation is more complicated. Let us mention only that it is conjectured that on projective spaces $\mathbb{P}^n$ , $n \geq 4$, almost all complex topological vector bundles do not have holomorphic structures, but it is not known even one such example ! For more results see, for example, [O-S-S, B-P].

Secondly, we present the *classification problem* for holomorphic vector bundles over a compact complex manifold $X$. By Proposition 1.3, we know that the isomorphism classes of holomorphic vector bundles of rank $r$ ($\mathrm{Vect}_{hol}^r(X)$) are described by the cohomology set $H^1(X, GL(r, \mathcal{O}_X))$. Naively, we can ask the question of the computation of this cohomology set in order to describe all holomorphic vector bundles of fixed rank. We shall see in Chapter 3 that, even in the case of line bundles (where it means the computation of the Picard group), the computation is difficult and it can be done in few cases of compact complex manifolds. Now we shall refine the question. Let $E$ be a fixed topological (smooth) vector bundle over $X$. We can ask to classify all holomorphic vector bundles in $E$ and, of course, we would like these holomorphic structures to form a complex space. Unfortunately, "jumping structures" (see [Bn, St1]) prohibit the existence of such moduli spaces. Thus we have to refine again the question. Let $P$ be a property referring to holomorphic vector bundles (simple, stable, semi-stable, Hermite-Einstein, etc.).

**Problem II (classification)** *Is it possible to endow the set of isomorphism classes of $P$-holomorphic structures in $E$ with a complex space structure ?*

We need more precise definitions.

**Definition 1.39** A *family of $P$-holomorphic structures* in $E$ parametrised by a complex space $S$ is a holomorphic vector bundle $\mathcal{E}$ over the product $S \times X$ such that

(1) the restriction $\mathcal{E}_s$ of $\mathcal{E}$ to the fibre $X_s = \{s\} \times X$ for every $s \in S$ is topologically (smoothly) isomorphic to $E$;

(2) $\mathcal{E}_s$ has property $P$ for every point $s \in S$.

Two such families $\mathcal{E}$ , $\mathcal{E}'$ are called *equivalent* if there exists a line bundle $\mathcal{L} \in \text{Pic}(S)$ such that $\mathcal{E}' \cong \mathcal{E} \otimes p^*(\mathcal{L})$, where $p : S \times X \to S$ is the first projection.

Let $\mathcal{A}n$ denotes the category of complex spaces and let

$$M_X^{P,E} : \mathcal{A}n \longrightarrow \text{Sets}$$

be the functor which associates to a complex space $S$ the set of equivalence classes of families of $P$-holomorphic structures in $E$ parametrised by $S$. Clearly, for a morphism of complex spaces $f : T \to S$ we have that

$$M_X^{P,E}(f) : M_X^{P,E}(S) \longrightarrow M_X^{P,E}(T)$$

is given by $[\mathcal{E}] \mapsto [f^*(\mathcal{E})]$, where $[\mathcal{E}]$ is the equivalence class of $\mathcal{E}$.

**Definition 1.40** Let $F : \mathcal{A}n \longrightarrow \text{Sets}$ be a contravariant functor. Then $F$ possesses a *fine moduli space* if $F$ is representable, i.e. if there exists a complex space $Y$ in $\mathcal{A}n$ together with an isomorphism $F \to h_Y := \text{Hom}(\cdot, Y)$ of functors.

For the following definition see [K-O]:

**Definition 1.41** We say that the contravariant functor $F : \mathcal{A}n \longrightarrow \text{Sets}$ has a *coarse moduli space*, if the covariant functor $\tilde{F} : \mathcal{A}n \longrightarrow \text{Sets}$, $\tilde{F}(S) := \text{Hom}(F, h_S)$, is representable by a complex space $Z$ such that the canonical mapping $F(\text{Spec}\mathbb{C}) \to |Z|$ is bijective (where $|Z|$ is the support-set of the complex space $Z$).

Obviously, a fine moduli space is also a coarse one. Such a complex space, if it exists for the functor $M_X^{P,E}$, is unique up to isomorphism, and is called a *coarse moduli space* (respectively, a *fine moduli space*) for $P$-holomorphic vector bundles over $X$ with topological (smooth) support $E$. Therefore, the classification problem II can be formulated in the following more precise form:

**Problem III (classification)** *Does there exist at least a coarse moduli space for the functor $M_X^{P,E}$ ?*

We shall present in Chapter 5 results on this theme. Let us only mention here that for simple holomorphic structures in a fixed smooth vector bundle $E$ over a compact complex manifold the answer is positive; see [L-O1, K-O, Mj, Nr2].

For vector bundles over curves, see [At2, Do1, Gk1, H-N, N-S, Nw, Tj1].

We end this section with some remarks on moduli spaces for the functor $M_X^{P,E}$. Let us suppose that there exists a fine moduli space denoted $\mathcal{M}_X^{P,E}$. From the definition we get an isomorphism

$$M_X^{P,E}(\mathcal{M}_X^{P,E}) \xrightarrow{\sim} \operatorname{Hom}(\mathcal{M}_X^{P,E}, \mathcal{M}_X^{P,E}) \,.$$

Let $\mathcal{E}_{\mathcal{M}}$ be a holomorphic vector bundle over the product $\mathcal{M}_X^{P,E} \times X$, whose isomorphism class corresponds under the above isomorphism to the identity morphism of $\mathcal{M}_X^{P,E}$. Then, for every complex space $S$ we get from the isomorphism

$$M_X^{P,E}(S) \xrightarrow{\sim} \operatorname{Hom}(S, \mathcal{M}_X^{P,E})$$

that for every family $\mathcal{E}$ of $P$-holomorphic structures in $E$ parametrised by $S$ there exists a unique morphism $u : S \to \mathcal{M}_X^{P,E}$ such that

$$\mathcal{E} \cong (u \times 1_X)^*(\mathcal{E}_{\mathcal{M}}) \,.$$

The vector bundle $\mathcal{E}_{\mathcal{M}}$ is called the (global) *universal bundle* over $\mathcal{M}_X^{P,E} \times X$. In particular, for $S = \operatorname{Spec}\mathbb{C}$ we get the bijection

$$M_X^{P,E}(\operatorname{Spec}\mathbb{C}) \simeq |\mathcal{M}_X^{P,E}| \,;$$

i.e. the points of the complex space $\mathcal{M}_X^{P,E}$ are in bijection with the isomorphism classes of $P$-holomorphic structures in $E$. But this is exactly the positive answer to the classification problem II.

Let us suppose now that there exists only a coarse moduli space also denoted by $\mathcal{M}_X^{P,E}$. From the definition we get also the bijection $M_X^{P,E}(\operatorname{Spec}\mathbb{C}) \simeq |\mathcal{M}_X^{P,E}|$ (i.e. the positive answer to the classification problem II) and the isomorphism:

$$\operatorname{Hom}(M_X^{P,E}, h_{\mathcal{M}_X^{P,E}}) \xrightarrow{\sim} \operatorname{Hom}(\mathcal{M}_X^{P,E}, \mathcal{M}_X^{P,E}) \,.$$

It follows that there exists a morphism of functors

$$\varphi : M_X^{P,E}(\cdot) \longrightarrow \operatorname{Hom}(\cdot, \mathcal{M}_X^{P,E})$$

such that for every space $S$ and family class $[\mathcal{E}] \in M_X^{P,E}(S)$, we have $\varphi(\mathcal{E})(s) = [\mathcal{E}_s]$, for all points $s \in S$, where we identified $|\mathcal{M}_X^{P,E}|$ with $M_X^{P,E}(\mathrm{Spec}\mathbb{C})$. In general there exist locally universal bundles. If a (global) universal bundle exists, then $\mathcal{M}^{P,E}$ represents the functor $M_X^{P,E}$ and it is a fine moduli space.

# 2. Facts on compact complex surfaces

In this chapter we present some facts on line bundles and divisors on compact complex manifolds and we define the Kodaira dimension and the algebraic dimension. Then we give (following essentially [B-P-V]) an outline of the Enriques-Kodaira classification of compact complex surfaces as well as examples in each class of surfaces with emphasis to the nonalgebraic case. Some results, which we explicitly need later, are presented for the convenience of the reader with more details. Finally, we precise the properties of the quadratic intersection form on the Neron-Severi group of a nonalgebraic surface.

## 2.1 Line bundles and divisors

Let $X$ be a complex manifold and let $\mathcal{O}_X$ denote its structural sheaf. We recall that the set of isomorphism classes of holomorphic line bundles over $X$ forms a group, which is naturally isomorphic to $H^1(X, \mathcal{O}_X^*)$, and it is called the *Picard group*, denoted by $\text{Pic}(X)$;( Section 1.1). From the exponential sequence of sheaves

$$0 \to \mathbb{Z}_X \to \mathcal{O}_X \overset{\exp}{\to} \mathcal{O}_X^* \to 0$$

we get the exact cohomology sequence

$$\to H^1(X, \mathbb{Z}) \to H^1(X, \mathcal{O}_X) \to H^1(X, \mathcal{O}_X^*) \overset{\delta}{\to} H^2(X, \mathbb{Z}) \to H^2(X, \mathcal{O}_X) \to,$$

where $\delta(\text{cls}(L)) = c_1(L)$ (Sect.1.2). If we put $\text{Pic}_0(X) := \text{Ker } \delta$, then $\text{Pic}(X)/\text{Pic}_0(X)$ is isomorphic to a subgroup $\text{Im } \delta$ of $H^2(X, \mathbb{Z})$, the *Neron-Severi group* of $X$, denoted by $\text{NS}(X)$. If $X$ is compact, then $H^2(X, \mathbb{Z})$ has finite rank; it follows that $\text{NS}(X)$ has finite rank $\rho(X)$, called the *Picard number* of $X$.

**Example** If $X = \mathbb{P}^n$, then $H^2(X, \mathbb{Z}) \cong \mathbb{Z}$ and $H^i(X, \mathcal{O}_X) = 0$ for $i > 0$. It follows that $\text{Pic}(\mathbb{P}^n) \cong \mathbb{Z}$ and a generator is $\mathcal{O}_{\mathbb{P}^n}(1)$. Thus, on $\mathbb{P}^n$ every holomorphic line bundle is isomorphic to $\mathcal{O}_{\mathbb{P}^n}(k)$ for a suitable integer $k$.

A *hypersurface* $H \subset X$ is any (non-empty) closed subset of $X$ with the property that every point $x \in H$ has a connected neighbourhood $U$, such that $H \cap U$ is the zero-set of a non-constant holomorphic function on $U$. A hypersurface is irreducible if it is not the union of two other hypersurfaces. A *Weil divisor* on $X$ is a formal sum $D = \sum_{i=1}^{\infty} n_i D_i$, where $n_i \in \mathbb{Z}$ and $(D_i)_i$ is a locally finite sequence of irreducible hypersurfaces on $X$ (locally finite means that every point has a neighbourhood which meets only finitely many $D_i$'s). The divisor $D$ is called *effective* ($D > 0$) if all $n_i$ are non-negative and not all zero. The Weil divisors on $X$ form an abelian group denoted by $\mathrm{Div}(X)$. If $X$ is compact, then $\mathrm{Div}(X)$ is the free abelian group generated by the irreducible hypersurfaces of the manifold $X$.

We shall show now, how to associate a line bundle to every hypersurface. Let $H \subset X$ be a hypersurface. We can find an open covering $\mathcal{U} = (U_i)$ of $X$ and holomorphic functions $f_i$ on $U_i$ such that each $f_i$ generates on $U_i$ the ideal sheaf $\mathcal{I}_H$ (if $H$ is a submanifold, then $f_i$ is a local equation for $H$ on $U_i$). On the intersections $U_i \cap U_j$ the functions $g_{ij} := f_i/f_j$ are holomorphic and invertible (sections of the sheaf $\mathcal{O}_X^*$). Then, by Section 1.1, they define a line bundle denoted by $\mathcal{O}_X(H)$. Now, by linearity, we can associate to every divisor $D$ a line bundle $\mathcal{O}_X(D)$ (the formal sum is finite !) and we get a linear map $\mathrm{Div}(X) \to \mathrm{Pic}(X)$.

Let us denote by $\mathcal{M}$ the sheaf of meromorphic functions on $X$ and by $\mathcal{M}^*$ the sheaf of invertible meromorphic functions. From the inclusion $\mathcal{O}_X \subset \mathcal{M}$ we get an exact sequence of sheaves:

$$1 \to \mathcal{O}_X^* \to \mathcal{M}^* \to \mathcal{M}^*/\mathcal{O}_X^* \to 1 .$$

Any global section of the sheaf $\mathcal{M}^*/\mathcal{O}_X^*$ is called a *Cartier divisor* on $X$ and $H^0(X, \mathcal{M}^*/\mathcal{O}_X^*)$ is called the *group of Cartier divisors* on $X$. If $s$ is a meromorphic function defined on an open set $U \subset X$, then we can associate to it a Weil divisor $(s)$ on $U$, by taking the zeros and the poles of $s$. If $f$ is a holomorphic and invertible function on $U$, then $(fs) = (s)$. From these facts we deduce that there is a linear map

$$H^0(X, \mathcal{M}^*/\mathcal{O}_X^*) \longrightarrow \mathrm{Div}(X) ,$$

from the group of Cartier divisors to the group of Weil divisors. By Weil [We], we know that this map is an isomorphism. Taking the cohomology we get the exact sequence

$$H^0(X, \mathcal{M}^*) \to \mathrm{Div}(X) \to H^1(X, \mathcal{O}_X^*) \to H^1(X, \mathcal{M}^*) . \tag{2.1}$$

By identifying $H^1(X, \mathcal{O}_X^*)$ with $\mathrm{Pic}(X)$, the homomorphism

$$\mathrm{Div}(X) \to H^1(X, \mathcal{O}_X^*)$$

coincides with the homomorphism $\mathrm{Div}(X) \to \mathrm{Pic}(X)$ defined above. The image of the homomorphism $H^0(X, \mathcal{M}^*) \to \mathrm{Div}(X)$ is called the *group of principal divisors* and it is denoted by $\mathrm{Div}_p(X)$ (a principal divisor has the form $(s)$, where $s$ is a global meromorphic function). We get an injective map

$$\mathrm{Div}(X)/\mathrm{Div}_p(X) \hookrightarrow \mathrm{Pic}(X) ,$$

where $\mathrm{Div}(X)/\mathrm{Div}_p(X)$ is called the *group of linear equivalence of divisors* on $X$. We have the following result:

**Proposition 2.1** *For a projective manifold $X$ we have the isomorphism*

$$Div(X)/Div_p(X) \xrightarrow{\sim} Pic(X) .$$

By a result of Lefschetz the algebraic Picard group is isomorphic to the analytic Picard group, i.e. the natural homomorphism

$$H^1(X_{alg}, \mathcal{O}^*_{X,alg}) \longrightarrow H^1(X, \mathcal{O}^*_X)$$

is an isomorphism. Then we compute the Picard group algebraically. Since $\mathcal{M}^*_{alg}$ is a constant sheaf in the Zariski topology and since $X$ is irreducible (any two Zariski open subsets have a non-empty intersection) it follows that $H^1(X_{alg}, \mathcal{M}^*_{alg}) = 0$. But $H^0(X_{alg}, \mathcal{M}_{alg}) = H^0(X, \mathcal{M})$ and any analytic hypersurface is an algebraic hypersurface (by GAGA). By the algebraic exact sequence analogous to (2.1) we get the desired isomorphism

$$\mathrm{Div}(X)/\mathrm{Div}_p(X) \xrightarrow{\sim} \mathrm{Pic}(X) \,.$$

*Remark.* It follows by the above result that on projective manifolds it does not matter if we work with line bundles or linear equivalence classes of divisors. In the nonalgebraic case, generally, there is no such isomorphism.

Let us recall the following important result (see [B-P-V], p. 45 , [Hh1], p. 179):

**Theorem 2.2 (Bertini)** *Let $X$ be a connected $n$-dimensional algebraic submanifold of $\mathbb{P}^N$, with $n \geq 2$. Then a general hyperplane $H$ in $\mathbb{P}^N$ intersects $X$ transversally along a smooth irreducible divisor.*

Thus, on projective manifolds there are "many divisors". The next example we shall give will present a completely opposite situation. It will be a nonalgebraic surface without divisors (curves). By a complex surface we shall mean a connected compact complex manifold of complex dimension two. Recall that a 2-torus is a quotient of $\mathbb{C}^2$ by a lattice $\Gamma$ of real rank 4 (i.e. $\Gamma$ is a free subgroup of $\mathbb{C}^2$ of rank 4). For the following example see [Sf2], Chapter VIII:

**Example 2.3** Let $\Gamma$ be defined by the vectors $v_1, v_2, v_3, v_4$ linearly independent over $\mathbb{R}$,

$$\Gamma = \mathbb{Z}v_1 + \mathbb{Z}v_2 + \mathbb{Z}v_3 + \mathbb{Z}v_4$$

and let $X = \mathbb{C}^2/\Gamma$. The complex surface $X$ is homeomorphic to $(\mathbb{R}/\mathbb{Z})^4$ and $H_2(X, \mathbb{Z}) \simeq \mathbb{Z}^6$, where the group $H_2(X, \mathbb{Z})$ is generated by the cycles $S_{ij}$ , $1 \leq i < j \leq 4$, obtained as the images of the planes $\mathbb{R}v_i + \mathbb{R}v_j$ by the natural projection $\mathbb{C}^2 \to X$. Let $C \subset X$ be a curve on $X$. Taking a triangulation of the normalization of $C$ we can consider the curve $C$ in a natural way as a singular cycle on $X$. In homology we have

$$C \sim \sum_{i<j} a_{ij} S_{ij} \,, \ a_{ij} \in \mathbb{Z} \,.$$

We shall prove that $C$ is not homologous to 0 (in fact, this is a general property of analytic submanifolds of a Kähler manifold !). The differential form on $\mathbb{C}^2$

$$\omega_1 = \frac{1}{2i}(\mathrm{d}z_1 \wedge \mathrm{d}\bar{z}_1 + \mathrm{d}z_2 \wedge \mathrm{d}\bar{z}_2)$$

induces a differential form $\omega$ on $X$ with the property

$$\int_C \omega > 0 .$$

It follows that not all the coefficients $a_{ij}$ are zero, therefore $C$ is not homologous to $0$.

Let now $\eta$ be the differential form on $X$ induced by $dz_1 \wedge dz_2$. For $v_i = (\alpha_i, \beta_i)$ we get by direct computation:

$$\int_{S_{ij}} \eta = \alpha_i \beta_j - \alpha_j \beta_i \ , \quad \int_C \eta = 0 .$$

On the other hand, by Stokes formula we have:

$$\int_C \eta = \sum_{i<j} a_{ij} \int_{S_{ij}} \eta .$$

Thus, if there exists a curve $C \subset X$, we get the relation:

$$\int_C \eta = \sum_{i<j} a_{ij}(\alpha_i \beta_j - \alpha_j \beta_i) = 0 , \qquad (2.2)$$

where not all the coefficients $a_{ij}$ are zero.

By a suitable choice of the real numbers $\alpha_1, \cdots, \alpha_4, \beta_1, \cdots, \beta_4$, we can obtain the fact that the relation (2.2) implies $a_{ij} = 0$ , $1 \leq i < j \leq 4$. It follows that the corresponding 2-torus has no curves (and is not algebraic), i.e. $\mathrm{Div}(X) = 0$. But the Picard group $\mathrm{Pic}(X)$ is big since it contains $\mathrm{Pic}_0(X)$, the dual torus $\mathbb{C}^2/\Gamma^*$ ($\Gamma^*$ is the dual lattice).

Now, let $Y$ be a $(n-1)$-dimensional complex submanifold of the complex manifold $X$. Obviously, $Y$ is a divisor on $X$. By the Example (6) in Section 1.1 we have

$$\mathcal{O}_X(Y)|_Y \cong N_{Y/X} ,$$

where $|$ denotes the analytic restriction. From the exact sequence

$$0 \to T_Y \to T_X|_Y \to N_{Y/X} \to 0$$

we get the isomorphism

$$\bigwedge^n T_X^*|_Y \cong \bigwedge^{n-1} T_Y^* \otimes N_{Y/X}^* .$$

We obtain

**Theorem 2.4 (adjunction formula)** *If $Y$ is a complex submanifold of codimension 1 of the complex manifold $X$, then*

$$\mathcal{K}_Y \cong \mathcal{K}_X \otimes \mathcal{O}_X(Y)|_Y .$$

Let now $X$ be a compact complex (nonsingular) surface. Then

$$H^4(X, \mathbb{Z}) \cong H_0(X, \mathbb{Z}) \cong \mathbb{Z}$$

in a canonical way, and we have the cup-product pairing

$$H^2(X, \mathbb{Z}) \times H^2(X, \mathbb{Z}) \longrightarrow H^4(X, \mathbb{Z}) \cong \mathbb{Z}$$

$$(\xi, \eta) \mapsto \xi.\eta \ .$$

For any two line bundles $L, M$ on $X$ we set

$$L.M := c_1(L).c_1(M) \ ,$$

and we call all the integers thus defined *intersection numbers*. If $D$ and $E$ are two divisors on $X$ we shall denote

$$D.E := \mathcal{O}_X(D).\mathcal{O}_X(E) = c_1(\mathcal{O}_X(D)).c_1(\mathcal{O}_X(E)) \ .$$

Let $x \in X$ be an isolated intersection point of two reduced curves $C$ and $D$ on the (nonsingular) surface $X$. Let $f, g \in \mathcal{O}_{X,x}$ be local equations for $C, D$ respectively. Then we can define the number

$$i_x(C, D) := \dim_{\mathbb{C}} \mathcal{O}_{X,x}/(f, g) \ .$$

We have (see [B-P-V], p. 66, [Hh1], p. 357):

**Theorem 2.5** *The intersection product thus defined has the following properties:*

*(1) It is bilinear with respect to the tensor product operation on line bundles and with respect to addition of divisors.*

*(2) It is symmetric.*

*(3) If $\pi : Y \to X$ is a proper map from another surface $Y$ onto $X$, then*

$$\pi^*(L).\pi^*(M) = (deg \ \pi)L.M \ .$$

*(4) If $C$ is any curve on $X$, then*

$$C.M = deg(M|_C) \ .$$

*(5) If $D_1$ and $D_2$ are two divisors, then $D_1.D_2$ is the intersection product of their homology classes.*

*(6) If the divisors $D, E$ have no common component (i.e. if they intersect in a finite set of points), then*

$$D.E = \sum_{x \in D \cap E} i_x(D, E) \ .$$

## 2.2 Algebraic dimension and Kodaira dimension

Let $X$ be a connected compact complex manifold and let us denote by $\mathcal{M}$ the sheaf of meromorphic functions on $X$. Then $H^0(X, \mathcal{M})$ is the field of (global) meromorphic functions on $X$ and we have (see [Sg]):

**Theorem 2.6 (Siegel)** *The field of meromorphic functions on a connected compact complex manifold $X$ is finitely generated over $\mathbb{C}$ and the transcendency degree over $\mathbb{C}$ does not exceed the dimension of $X$.*

The transcendency degree is called the *algebraic dimension* of $X$ and will be denoted by $a(X)$. It follows from GAGA (Section 1.3) that if $X$ is a projective manifold, then every meromorphic function is rational and the field of meromorphic functions is nothing but the field of rational functions on $X$.

By a complex surface we mean a connected compact complex manifold of complex dimension two. For a complex surface the algebraic dimension $a(X)$ can take the values $0, 1$ or $2$. A complex surface is called *algebraic* if it has an embedding in a projective space $\mathbb{P}^n$ (recall that by the Chow Theorem (Corollary 1.15) any complex submanifold of a projective space is algebraic and, that any abstract (nonsingular) compact algebraic surface is projective). Clearly, for the 2-torus in the Example 2.3 we have $a(X) = 0$. The condition $a(X) = 2$ characterizes the algebraic surfaces (see [C-K]):

**Theorem 2.7 (Chow-Kodaira)** *A complex surface $X$ is algebraic if and only if its algebraic dimension $a(X) = 2$.*

Recall also the following fundamental result (see [Kd1]):

**Theorem 2.8 (Kodaira)** *A complex surface $X$ is algebraic if and only if there exists over $X$ a holomorphic line bundle $L$ such that $c_1^2(L) > 0$.*

It follows

**Corollary 2.9** *If $X$ is a nonalgebraic surface, then $c_1^2(L) \leq 0$ for any holomorphic line bundle $L$ over $X$. (i.e. the semi-negativity of the intersection form on the Neron-Severi group $NS(X)$).*

**Corollary 2.10** *If $X$ is a nonalgebraic surface and $L \in Pic(X)$ such that $c_1^2(L) = 0$, then $c_1(L).c_1(M) = 0$ for any $M \in Pic(X)$ (i.e. $c_1(L)$ is orthogonal on the Neron-Severi group $NS(X)$).*

*Proof.* We have

$$c_1^2(L^{\otimes m} \otimes M) = 2m c_1(L).c_1(M) + c_1^2(M) .$$

If $c_1(L).c_1(M) \neq 0$ then, for a suitable integer $m$, it follows $c_1^2(L^{\otimes m} \otimes M) > 0$ and by the above theorem we get that $X$ is algebraic, contradiction. Therefore $c_1(L).c_1(M) = 0$.

**Definition 2.11** A connected holomorphic map $f : X \to S$ from a surface $X$ onto a nonsingular curve $S$ is called an *elliptic fibration* if the general fibre $X_s$ , $s \in S$, is a smooth elliptic curve. A surface is called *elliptic* if it admits some elliptic fibration.

**Corollary 2.12** *Any connected holomorphic map from a nonalgebraic surface $X$ onto a nonsingular curve $S$ is an elliptic fibration.*

*Proof.* The general fibre $C$ is nonsingular and $C^2 = 0$. The previous corollary now gives $\mathcal{K}_X.C = 0$, where $\mathcal{K}_X$ denotes the class of the canonical bundle over $X$. By adjunction formula (Theorem 2.4)

$$2g(C) - 2 = \mathcal{K}_X.C + C^2$$

we get $g(C) = 1$, i.e. $C$ is an elliptic curve.

**Theorem 2.13** *For $X$ a nonalgebraic surface we have $a(X) = 1$ if and only if $X$ is elliptic.*

*Proof.* If $a(X) = 1$ there exists a non-constant meromorphic function $h$ on $X$. If $h$ had indeterminacy points we would have $D^2 > 0$, where $D = (h)$ is the associated divisor of $h$. Since $X$ is nonalgebraic, this cannot be the case, hence $h$ defines a morphism (still denoted by $h$) $h : X \to \mathbb{P}^1$. Its Stein factorization (see [Fh], p. 71) gives a nonsingular curve $S$ and morphisms $f : X \to S$ , $g : S \to \mathbb{P}^1$ such that $f$ is connected and $h = g \circ f$. By the above corollary $f : X \to S$ is an elliptic fibration, hence $X$ is elliptic.

Conversely, if $X$ is elliptic, then there exists an elliptic fibration $f : X \to S$ and we have $f^*(H^0(S, \mathcal{M}_S)) \subset H^0(X, \mathcal{M}_X)$. Since $S$ is a curve $\dim_{\mathbb{C}} H^0(S, \mathcal{M}_S) = 1$, hence $a(X) \geq 1$. Since $X$ was assumed nonalgebraic it follows $a(X) = 1$ (in fact, it follows the equality $f^*(H^0(S, \mathcal{M}_S)) = H^0(X, \mathcal{M}_X)$).

The algebraic dimension of a surface $X$ can be expressed in terms of "number of curves" on $X$. When $a(X) = 2$, $X$ is projective hence, by Bertini Theorem, through each point we have an infinity of curves on $X$. This is no longer true when $X$ is nonalgebraic (i.e. $a(X) \leq 1$) as we shall see in what follows.

**Proposition 2.14** *Let $X$ be a nonalgebraic surface with $a(X) = 1$ and let $f : X \to S$ be an elliptic fibration. Then any irreducible curve on $X$ is contained in some fibre and thus the fibration is unique.*

*Proof.* Let $D$ be an irreducible curve contained in no fibre and let $x_0 \in D$ be a point on $D$. Let $s = f(x_0) \in S$ and let $C = f^{-1}(s)$; then $C.D > 0$. But $C^2 = 0$ and we get a contradiction by Corollary 2.10.

Let $X$ be a compact surface. If there exist on $X$ two linearly independent holomorphic 1-forms $\omega_1, \omega_2$ such that $\omega_1 \wedge \omega_2 \equiv 0$, then there exists a nonsingular curve $S$ and a connected holomorphic map $f : X \to S$, such that $\omega_1 = f^*(\alpha_1)$ , $\omega_2 = f^*(\alpha_2)$ with $\alpha_1$ , $\alpha_2$ holomorphic 1-forms on $S$ (see [B-P-V], Chap.IV , Prop. 4.1). In particular, it follows $a(X) \geq 1$.

**Proposition 2.15** *Let $X$ be a nonalgebraic surface with $a(X) = 0$. Then:*

*(1)* $h^0(X, L) \leq 1$ *for any line bundle* $L \in Pic(X)$; *in particular* $p_g(X) := h^0(X, \Omega_X^2) \leq 1$;

*(2)* $h^{1,0}(X) := \dim_{\mathbb{C}} H^0(X, \Omega_X^1) \leq 2$.

*Proof.* (1) If $s_1$ and $s_2$ are two linearly independent (over $\mathbb{C}$) sections of the line bundle $L$, then $s_1/s_2$ is a (global) meromorphic function on $X$ which is not constant. It follows $a(X) \geq 1$, contradiction.

(2) Let $\omega_1, \omega_2$ and $\omega_3$ be three linearly independent holomorphic 1-forms on $X$. Then $\omega_1 \wedge \omega_2$ and $\omega_1 \wedge \omega_3$ are not identically zero on $X$, otherwise it would follow $a(X) \geq 1$ (see the previous remark). By (1) we have

$$\lambda \omega_1 \wedge \omega_2 + \mu \omega_1 \wedge \omega_3 \equiv 0$$

for some non-zero constants $\lambda, \mu \in \mathbb{C}$. It follows $\omega_1 \wedge (\lambda \omega_2 + \mu \omega_3) \equiv 0$ and, again, we get $a(X) \geq 1$, contradiction. Thus $h^{1,0}(X) \leq 2$.

**Theorem 2.16** *If $X$ is a nonalgebraic surface having $a(X) = 0$, then the number of irreducible curves on $X$ is finite, not exceeding $h^{1,1}(X) + 2$.*

*Proof.* (cf. [F-F]) Let $\mathcal{M}^1$, $\Omega^1$ be the sheaves of germs of meromorphic, resp. holomorphic 1-forms on $X$ and let $\mathcal{L} := \mathcal{M}^1/\Omega^1$. A curve on $X$ of local equation $f = 0$ defines an element $\frac{df}{f} \in H^0(X, \mathcal{L})$. One can see that different irreducible curves give linearly independent elements in $H^0(X, \mathcal{L})$. One considers then the long exact cohomology sequence

$$\cdots \to H^0(X, \mathcal{M}^1) \to H^0(X, \mathcal{L}) \to H^1(X, \Omega^1) \to \cdots$$

and we only need to prove $h^0(X, \mathcal{M}^1) \leq 2$. Let $\omega_1, \omega_2 \in H^0(X, \mathcal{M}^1)$ be linearly independent over $\mathbb{C}$. Locally on $U \subset X$ they are linearly independent over $\mathcal{M}(U)$, otherwise $\omega_1 \wedge \omega_2 = 0$ on $U$, hence on $X$ and $\omega_1/\omega_2 \in \mathcal{M}(X) = \mathbb{C}$ $(a(X) = 0)$. Thus, locally on $U$ every third form $\omega \in \mathcal{M}^1(X)$ can be uniquely written as a linear combination in $\omega_1, \omega_2$ over $\mathcal{M}(U)$ which globalizes to give

$$\omega = f_1 \omega_1 + f_2 \omega_2 \quad \text{with} \quad f_1, f_2 \in \mathcal{M}(X) = \mathbb{C} .$$

It follows that the number of irreducible curves on $X$ is finite, not exceeding

$$\dim_{\mathbb{C}} H^0(\mathcal{L}) \leq h^{1,1}(X) + 2 .$$

In order to describe the classification theorem for compact complex surfaces we recall the important concepts of Kodaira dimension and minimal model. For details and proofs see [Bd, B-P-V, B-H, Mm2, Uel].

Let $X$ be any compact complex manifold and let $\mathcal{K}_X$ be the canonical line bundle. Since there is a pairing

$$H^0(X, \mathcal{K}_X^{\otimes m_1}) \otimes H^0(X, \mathcal{K}_X^{\otimes m_2}) \longrightarrow H^0(X, \mathcal{K}_X^{\otimes(m_1+m_2)})$$

we can make the direct sum

$$R(X) := \mathbb{C} \oplus \sum_{m \geq 1} H^0(X, \mathcal{K}_X^{\otimes m})$$

into a commutative ring $R(X)$ with unit element. This ring is called the *canonical ring* of $X$. It can be proved that $R(X)$ has a finite degree of transcendency, say $\mathrm{tr}(R(X))$, over $\mathbb{C}$ (see [Ue1]). Thus we have the following

**Definition 2.17** The *Kodaira dimension* of the compact complex manifold $X$ is

$$\mathrm{kod}(X) := \begin{cases} -\infty & \text{if } R(X) \cong \mathbb{C} \\ \mathrm{tr}(R(X)) - 1 & \text{otherwise.} \end{cases}$$

We have always that

$$\mathrm{kod}(X) \leq a(X) \leq \dim X .$$

Let $P_m(X) := h^0(X, \mathcal{K}_X^{\otimes m})$ , $m \geq 1$. This number is called the *m-th plurigenus* of $X$ $(P_1(X) = p_g(X))$. The Kodaira dimension yields precise information about the behaviour of $P_m(X)$ for $m \to \infty$. We have (see [Ue1], p. 86, [Ue2]):

**Theorem 2.18** *Let $X$ be a compact complex manifold. Then:*

$\mathrm{kod}(X) = -\infty \iff P_m(X) = 0$ *for all $m \geq 1$ ;*

$\mathrm{kod}(X) = 0 \iff P_m(X) = 0$ *or $1$ for $m \geq 1$, but not always $0$ ;*

$\mathrm{kod}(X) = k \iff$ *there exists an integer $k$ $(1 \leq k \leq \dim X)$ and strictly positive constants $\alpha, \beta$ such that*

$$\alpha m^k < P_m(X) < \beta m^k$$

*for $m$ large enough.*

*Remark.* For $\mathrm{kod}(X) = k \geq 1$ we have that $P_m(X)$ grows like $m^k$.

For the next result we refer again to [Ue1], p. 73:

**Theorem 2.19** *Let $X$ and $Y$ be compact, connected complex manifolds of the same dimension. If there exists a generically finite holomorphic map from $X$ onto $Y$, then $P_m(X) \geq P_m(Y)$ for $m \geq 1$, hence $\mathrm{kod}(X) \geq \mathrm{kod}(Y)$. If the map is an unramified covering, then $\mathrm{kod}(X) = \mathrm{kod}(Y)$.*

Now we shall present shortly the notions of $\sigma$-process (blow-up) and of contraction. Let $(z_1, \cdots, z_n)$ be the coordinates in $\mathbb{C}^n$ , $n \geq 2$, and let $(\xi_1 : \cdots : \xi_n)$ be the homogeneous coordinates in $\mathbb{P}^{n-1}$. We take a neighbourhood $U$ of a point $a = (a_1, \cdots, a_n)$ in $\mathbb{C}^n$ and consider on the product $U \times \mathbb{P}^{n-1}$ the subset $V$ given by the equations

$$(z_i - a_i)\xi_j - (z_j - a_j)\xi_i = 0 , \quad i, j = 1, \cdots, n .$$

It is easy to see that $V$ is an $n$-dimensional complex submanifold of $U \times \mathbb{P}^{n-1}$. The projection $p : V \to U$ maps $V \setminus p^{-1}(a)$ biregularly onto $U \setminus \{a\}$, whereas $p^{-1}(a)$ is an $(n-1)$-dimensional submanifold of $V$, isomorphic to $\mathbb{P}^{n-1}$. We shall call the map $p : V \to U$ $\sigma$-process (or *blow-up*) of $U$ in $a$. Using local coordinates we can blow up any point $x_0$ of an $n$-dimensional complex manifold $X$ ($n \geq 2$); up to an isomorphism the result is independent of the local coordinates used. In this way we can define a complex manifold $\tilde{X}$ and a natural map $p : \tilde{X} \to X$, called $\sigma$-process or *blow-up* of $X$ in $x_0$. It is clear that $E = p^{-1}(x_0)$ is a submanifold isomorphic to $\mathbb{P}^{n-1}$. The divisor $E$ will be called the *exceptional divisor*. By Levi's extension Theorem ( [Fh], p. 185) it follows that $p$ induces an isomorphism between the fields of meromorphic functions on $X$ and $\tilde{X}$. In particular, if $X$ (and hence $\tilde{X}$) is compact, then $a(X) = a(\tilde{X})$. Again by Levi's extension Theorem it follows that $p$ induces an isomorphism

$$p^* : H^0(X, \mathcal{K}_X^{\otimes m}) \xrightarrow{\sim} H^0(\tilde{X}. \mathcal{K}_{\tilde{X}}^{\otimes m})$$

for all $m \geq 1$, so if $X$ is compact, $P_m(\tilde{X}) = P_m(X)$ for $m \geq 1$ and $\mathrm{kod}(\tilde{X}) = \mathrm{kod}(X)$. Moreover, we have (see, for example, [B-P-V], p. 28):

**Theorem 2.20** *Let $X$ be a complex manifold of dimension $\geq 2$ and $p : \tilde{X} \to X$ the blow-up of $X$ at some point. Then:*

*(1)* $N_{E/\tilde{X}} \cong \mathcal{O}_{\mathbb{P}^{n-1}}(-1)$ ;

*(2)* $H^2(\tilde{X}, \mathbb{Z}) \cong H^2(X, \mathbb{Z}) \oplus \mathbb{Z}e$, *where* $e = c_1(\mathcal{O}_{\tilde{X}}(E))$ ;

*(3)* $\mathrm{Pic}(\tilde{X})$ *is isomorphic to the product of $\mathrm{Pic}(X)$ and the infinite cyclic subgroup generated by $\mathcal{O}_{\tilde{X}}(E)$* ;

*(4)* $\mathcal{K}_{\tilde{X}} \cong p^*(\mathcal{K}_X) \otimes \mathcal{O}_{\tilde{X}}((\dim X - 1)E)$.

In case $X$ is a surface, then $\tilde{X}$ is a surface, $E = p^{-1}(x_0)$ is a nonsingular rational curve and, moreover, $c_1^2(\mathcal{O}_{\tilde{X}}(E)) = E^2 = -1$.

Let $X$, $Y$ be irreducible reduced 2-dimensional complex spaces. A proper holomorphic surjective map $\pi : X \to Y$ is called *bimeromorphic* if there are proper analytic subsets $T \subset X$ and $S \subset Y$ such that $\pi : X \setminus T \to Y \setminus S$ is biholomorphic. A compact, reduced, connected curve $C$ on a nonsingular surface $X$ is called *exceptional*, if there is a bimeromorphic map $\pi : X \to Y$, such that there is an open neighbourhood $U$ of $C$ in $X$, a point $y \in Y$, and a neighbourhood $V$ of $y$ in $Y$ with the property that $\pi : U \setminus C \to V \setminus \{y\}$ is biholomorphic and $\pi(C) = y$. We shall say also in this case that $C$ is *contracted* to $y$. The following characterisation is due to Grauert [Gr2] and its algebraic version to Mumford [Mm1]:

**Theorem 2.21** *A reduced, compact, connected curve $C$ with irreducible components $C_i$ on a smooth surface is exceptional if and only if the intersection matrix $(C_i.C_j)$ is negative definite.*

The most important examples of exceptional curves are the *exceptional curves of the first kind*, i.e. nonsingular rational curves with self-intersection $-1$ ($C^2 = -1$), frequently called $(-1)$-*curves*. In fact, the contraction of a $(-1)$-curve is the inverse construction of a $\sigma$-process. We have the following result:

**Theorem 2.22** *Let $X$, $Y$ be nonsingular surfaces and $\pi : X \to Y$ a bimeromorphic map. Then $\pi$ is equivalent to a succession of $\sigma$-processes, which is locally finite with respect to $Y$.*

**Definition 2.23** A smooth surface is called *minimal*, if it does not contain any $(-1)$-curve. A nonsingular surface $X_{min}$ is called *minimal model* of the nonsingular surface $X$, if $X_{min}$ is minimal itself, and if there is a bimeromorphic map (i.e. a succession of $\sigma$-processes) from $X$ to $X_{min}$.

We have:

**Theorem 2.24** *Every compact nonsingular surface has a minimal model.*

We can argue as follows: Suppose that $X$ contains a $(-1)$-curve and let $X_1$ be obtain from $X$ by contracting it. If $X_1$ contains another $(-1)$-curve, the process can be repeated, and so on. This must lead to a surface without $(-1)$-curves after a finite number of blowing downs, since each time the second Betti number diminishes by 1 (Theorem 2.20).

One proves that if $X$ is a compact connected surface with $\mathrm{kod}(X) \geq 0$, then all minimal models of $X$ are isomorphic. If $\mathrm{kod}(X) = -\infty$, then $X$ could have non-isomorphic minimal models. This happens only in the algebraic case, in the nonalgebraic case the minimal model being unique.

## 2.3  Classification and examples of surfaces

Let $X$ be a complex surface. Then, the Kodaira dimension can take the values $\mathrm{kod}(X) = -\infty$, 0, 1 and 2. According to their Kodaira dimension, surfaces are divided into ten classes. This classification is called the Enriques-Kodaira classification (see [En, Kd1, Kd2, B-P-V, B-H, Bd, Be, G-H, Kr, Sfl]).

**Theorem 2.25 (Enriques-Kodaira)** *Every surface has a minimal model in exactly one of the classes (1) to (10) of the following Table 2.1. This model is unique (up to isomorphisms) except for the surfaces with minimal models in the classes (1) and (3).*

For a proof see [B-P-V], p. 187. We shall present the definitions and examples in all these classes (not necessarily in the order of the Table 2.1.) together with some additional results.

**Examples (1)** A *rational surface* is a surface that is birationally equivalent to $\mathbb{P}^2$. A *ruled surface* is a compact surface which admits a ruling, i.e. a compact surface which is the total space of an analytic fibre bundle with fibre $\mathbb{P}^1$ and structural group $PGL(2, \mathbb{C})$ over a smooth, connected curve $B$. Examples are provided by the projective bundles $\mathbb{P}(E)$ of algebraic 2-vector bundles $E$ over a smooth, connected, compact curve. In fact there are no other examples since any ruled surface is isomorphic to $\mathbb{P}(E)$, where $E$ is an algebraic 2-vector bundle over $B$. Since on $B = \mathbb{P}^1$, by a theorem of Grothendieck [Gk1], every algebraic vector bundle is isomorphic to a direct sum of line bundles and since $\mathbb{P}(E \otimes L) \cong \mathbb{P}(E)$ for any algebraic line bundle $L$ on $B$, it follows that every ruled surface over $\mathbb{P}^1$ is of the form $\mathbb{P}(\mathcal{O}_{\mathbb{P}^1} \oplus \mathcal{O}_{\mathbb{P}^1}(n))$ for some integer $n \geq 0$ ( $\mathcal{O}_{\mathbb{P}^k}(n)$ are all line bundles on $\mathbb{P}^k$ ; see Section 2.1). The surfaces $\mathbb{P}(\mathcal{O}_{\mathbb{P}^1} \oplus \mathcal{O}_{\mathbb{P}^1}(n))$ are denoted by $\Sigma_n$ and are called the *Hirzebruch surfaces*. They are birationally equivalent to $\mathbb{P}^1 \times \mathbb{P}^1$ hence to $\mathbb{P}^2$, so they are all rational. The surface $\Sigma_0$ is $\mathbb{P}^1 \times \mathbb{P}^1$, $\Sigma_1$ is $\mathbb{P}^2$ blown up in one point and also $\Sigma_n$ , $n \geq 2$, can be characterised in more geometric ways. One can show that $\mathbb{P}^2$ and the $\Sigma_n$ , $n \neq 1$, are the only minimal rational surfaces.

**(3)** Here we take ruled surfaces over a curve $B$ of genus $g \geq 1$. If $B$ is elliptic, then there are 2-vector bundles over $B$ which do not split (i.e. which are not direct sum of two line bundles). To see this we start from an extension

$$0 \to \mathcal{O}_B \overset{i}{\to} E \to \mathcal{O}_B \to 0$$

which does not split. These extensions are classified by $H^1(B, \mathcal{O}_B) \cong \mathbb{C}$ and it suffices to choose an extension which corresponds to a non-zero element in $\mathbb{C}$. Now we show that the 2-vector bundle $E$ does not split either. Let us assume that $E \cong L_1 \oplus L_2$, with $L_1$ , $L_2$ line bundles over $B$. Then $\mathcal{O}_B \cong \det(E) \cong L_1 \otimes L_2$. If, say $L_1$ were $i(\mathcal{O}_B)$, then $L_2$ would also be isomorphic to $\mathcal{O}_B$ and the extension would split. If neither $L_1$ nor $L_2$ were $i(\mathcal{O}_B)$ then, since both line bundles would admit non-trivial homomorphisms onto $\mathcal{O}_B$, we would have $L_1 \cong L_2 \cong \mathcal{O}_B$ and the extension would split again. Similarly, there exists an extension

$$0 \to \mathcal{O}_B \to F \to L \to 0 \ ,$$

such that the vector bundle $F$ does not split, where $L$ is some line bundle of degree 1. Atiyah has shown in [At2] that $\mathbb{P}(E)$ and $\mathbb{P}(F)$ are the only ruled surfaces over an elliptic curve $B$ which are not the projective bundle of a splitting vector bundle of rank 2. For the classification of ruled surfaces over a base of genus $\geq 2$ see [Tj1, Tj2].

**(2)** A *surface of class VII* is a surface $X$ with $\mathrm{kod}(X) = -\infty$ and $b_1(X) = 1$. These surfaces are nonalgebraic and even non-Kähler ($b_1 = 1$ !). We shall present some examples in this class.
Let $S^3$ be the sphere identified with $\{z = (z_1, z_2) \in \mathbb{C}^2 \mid |z_1|^2 + |z_2|^2 = 1\}$ and let $a \neq 1$ be a real positive number. One sees easily that the map

$$f_a : S^3 \times \mathbb{R} \longrightarrow \mathbb{C}^2 \setminus \{0\} \ , \ f_a(z_1, z_2, t) = (a^t z_1, a^t z_2) \ ,$$

is a diffeomorphism. The additive group $\mathbb{Z}$ acts (as a group of differentiable automorphisms) on $S^3 \times \mathbb{R}$ by

$$(z_1, z_2, t) \mapsto (z_1, z_2, t + m) \ , \ m \in \mathbb{Z} \ ,$$

and the corresponding quotient space $(S^3 \times \mathbb{R})/\mathbb{Z}$ is diffeomorphic to $S^3 \times S^1$. Through $f_a$ the action of the group $\mathbb{Z}$ on $S^3 \times \mathbb{R}$ can be view as an action of a group $\Gamma$ of analytic automorphisms on $\mathbb{C}^2 \setminus \{0\}$; in fact

**Table 2.1.**

| Nr. | class of $X$ | $kod(X)$ | $b_1(X)$ | $a(X)$ | order of $\mathcal{K}_X$ |
|-----|-------------|----------|----------|--------|---------------------------|
| (1) | minimal rational surfaces | $-\infty$ | 0 | 2 | |
| (2) | minimal surfaces of class VII | $-\infty$ | 1 | 0, 1 | |
| (3) | ruled surfaces of genus $g \geq 1$ | $-\infty$ | $2g$ | 2 | |
| (4) | Enriques surfaces | 0 | 0 | 2 | 2 |
| (5) | hyperelliptic surfaces | 0 | 2 | 2 | $2, 3, 4, 6$ |
| (6) | Kodaira surfaces | | | | |
| | (a) primary | 0 | 3 | 1 | 1 |
| | (b) secondary | 0 | 1 | 1 | $2, 3, 4, 6$ |
| (7) | K3-surfaces | 0 | 0 | $0, 1, 2$ | 1 |
| (8) | 2-tori | 0 | 4 | $0, 1, 2$ | 1 |
| (9) | minimal properly elliptic surfaces | 1 | | $1, 2$ | |
| (10) | minimal surfaces of general type | 2 | $\equiv 0(2)$ | 2 | |

$$(z_1, z_2, m) \mapsto (a^m z_1, a^m z_2) , \ m \in \mathbb{Z} , \ (z_1, z_2) \in \mathbb{C}^2 \setminus \{0\}.$$

Since this group $\Gamma$ acts properly discontinuous without fixed points, it follows that $H_a = (\mathbb{C}^2 \setminus \{0\})/\Gamma$ has a natural structure of a complex 2-dimensional manifold, called *Hopf surface*. But $H_a$ is diffeomorphic to $S^3 \times S^1$ so, in particular, it follows $b_1(H_a) = 1$, hence $H_a$ is not Kähler (and nonalgebraic). More generally, if $a = (a_1, a_2) \in \mathbb{C}^2$ such that $0 < |a_1| \leq |a_2| < 1$ then, by considering the group $\Gamma$ of analytic automorphisms of $\mathbb{C}^2 \setminus \{0\}$ defined by the action:

$$(z_1, z_2, m) \mapsto (a_1^m z_1, a_2^m z_2) , \ m \in \mathbb{Z} , \ (z_1, z_2) \in \mathbb{C}^2 \setminus \{0\},$$

we get a Hopf surface $H_a = (\mathbb{C}^2 \setminus \{0\})/\Gamma$. Again, $H_a$ is diffeomorphic to $S^3 \times S^1$ and hence it is not Kähler. We have the following result (see [B-P-V], p. 173):

**Proposition 2.26** *The Hopf surface $H_a$ is an elliptic fibre space over $\mathbb{P}^1$ (and $a(H_a) = 1$) if and only if $a_1^k = a_2^l$ for some $k, l \in \mathbb{Z}$. Otherwise, $H_a$ contains exactly two irreducible curves, the images of the punctured $z_1$-and $z_2$-axes (and $a(H_a) = 0$).*

More generally, a compact complex surface is called a *Hopf surface* if its universal covering is analytically isomorphic to $\mathbb{C}^2 \setminus \{0\}$. Kodaira has treated Hopf surfaces extensively (see [Kd1, Kd2]) and proved that every Hopf surface has Kodaira dimension $-\infty$. Let us mention one more result (see [Kd2]):

**Theorem 2.27** *(1) The minimal compact surfaces $X$ with $a(X) = 1$ , $kod(X) = -\infty$ are exactly the Hopf surfaces.*
*(2) A compact surface $X$ with $a(X) = 0$ is a Hopf surface if and only if $b_1(X) = 1$ , $b_2(X) = 0$ and there is a curve on $X$.*

Thus, besides some Hopf surfaces with $a(X) = 1$ all minimal surfaces in class VII (class (2) here) have $a(X) = 0$. For surfaces $X$ in class VII with $a(X) = 0$ and $b_2(X) = 0$, there are, apart from Hopf surfaces, other three kind of examples given by Inoue [In]. We shall present only one type of Inoue surfaces (see also [B-H]).

Let $M = (m_{ij}) \in SL(3, \mathbb{Z})$, and suppose that the eigenvalues of $M$ satisfy the following conditions: one of them, say $\alpha$, is real with $\alpha > 1$, and the other two, $\beta$ and $\bar{\beta}$, are not real ($\beta \neq \bar{\beta}$). For example, we can take for $M$ the matrix

$$\begin{pmatrix} 0 & 1 & 0 \\ n & 0 & 1 \\ 1 & 1-n & 0 \end{pmatrix}$$

with $n \in \mathbb{Z}$. By using the matrix $M$ we can define a compact complex surface in the following way. Let $(a_1, a_2, a_3)$ be a real eigenvector of $M$ corresponding to $\alpha$, and $(b_1, b_2, b_3)$ an eigenvector of $M$ corresponding to $\beta$. Since $(a_1, a_2, a_3)$, $(b_1, b_2, b_3)$ and $(\bar{b}_1, \bar{b}_2, \bar{b}_3)$ are independent over $\mathbb{C}$, the vectors $(a_1, b_1)$, $(a_2, b_2)$ and $(a_3, b_3)$ are independent over $\mathbb{R}$. Let $U$ be the upper half plane, and $G_M$ the group of analytic automorphisms of $U \times \mathbb{C}$, generated by $g_0, g_1, g_2, g_3$, where

$$\begin{aligned} g_0(w, z) &= (\alpha w, \beta z) \\ g_i(w, z) &= (w + a_i, z + b_i) \quad i = 1, 2, 3. \end{aligned}$$

One can prove (elementary) that $G_M$ acts properly and discontinuously without fix-points on $U \times \mathbb{C}$, so by [Cr] the quotient $S_M = U \times \mathbb{C}/G_M$ is a compact complex surface, called *Inoue surface*. One can also prove that $S_M$ has $b_1(S_M) = 1$, $b_2(S_M) = 0$, kod$(S_M) = -\infty$ and it does not contain any curve (hence $a(X) = 0$).

As we mentioned above there are other two types of Inoue surfaces with the same properties. In order to complete the list of class $VII_0$ (minimal) surfaces with $b_2 = 0$, it remains to describe the class

$$A = \{X \in VII_0 \mid b_2(X) = 0, \ X \text{ contains no curve}\}.$$

A very important step on this way was the following result (see [In]):

**Theorem 2.28 (Inoue)** *If a surface $X$ belonging to the class $A$ is not isomorphic to an Inoue surface, then its tangent bundle $T_X$ must be irreducible.*

For the definition of an irreducible vector bundle see Chapter 4. Then, the following result gives the complete classification for the class $A$ surfaces (see [Bol, L-Y-Z, Te2]).

**Theorem 2.29** *Every surface belonging to the class $A$ is isomorphic to an Inoue surface.*

*Remark.* It is interesting that the irreducibility of the tangent vector bundle $T_X$ was used through its stability with respect to a (any) Gauduchon metric by applying the Kobayashi-Hitchin correspondence (see Chapter 5 for precise definitions and statements).

Inoue and Hirzebruch constructed examples of surfaces in class VII with $b_2 > 0$ (Inoue-Hirzebruch surfaces). For more recent work in this direction see [Ek, Ka, Nk].

*Remark.* The classification of minimal surfaces $X$ in class VII with $a(X) = 0$ and $b_2(X) > 0$ is still open.

**(8)** A *2-torus* is a surface isomorphic to the quotient of $\mathbb{C}^2$ by a lattice $\Gamma$ of real rank 4 (see example 2.3). More generally, let $\Gamma \subset \mathbb{C}^n$ be a lattice generated by the vectors $\gamma_1, \cdots, \gamma_{2n} \in \mathbb{C}^n$, which are considered as column vectors. Define the $n \times 2n$ period matrix

$$\Pi := (\gamma_1 | \cdots | \gamma_{2n}) .$$

An *n-torus* is the quotient compact complex manifold $X = \mathbb{C}^n / \Gamma$. One has a natural isomorphism (topologically, $X \cong (S^1)^{2n}$):

$$H^2(X, \mathbb{Z}) \cong \mathrm{Alt}^2_{\mathbb{Z}}(\Gamma, \mathbb{Z})$$

of the cohomology group $H^2(X, \mathbb{Z})$ with the space of alternating integer-valued 2-forms on $\Gamma$. Let

$$H(\mathbb{C}^n, \Gamma) := \{ H \mid H \text{ hermitian form on } \mathbb{C}^n \text{ with } (\mathrm{Im}\ H)(\Gamma \times \Gamma) \subset \mathbb{Z} \} .$$

Since the imaginary part $\mathrm{Im}\ H$ of a hermitian form $H$ is an alternating 2-form which determines completely $H$, we may consider $H(\mathbb{C}^n, \Gamma)$ as a subgroup of $\mathrm{Alt}^2_{\mathbb{Z}}(\Gamma, \mathbb{Z}) \cong H^2(X, \mathbb{Z})$. With this identification one has by the Theorem of Appell-Humbert (see Mumford [Mm4], p. 20)

$$\mathrm{NS}(X) \cong H(\mathbb{C}^n, \Gamma) .$$

Let us call *Riemann form* on $X$ any hermitian form $H \in H(\mathbb{C}^n, \Gamma)$ which is positive semi-definite. The algebraic dimension of $X$ is given by (see [We])

$$a(X) = \max \{ \mathrm{rank}\ H \mid H \text{ Riemann form of } X \} .$$

Let us come back to the case of 2-tori. By an analytic isomorphism of the 2-torus $X$ we can choose the lattice $\Gamma$ to be generated by the column vectors of the following period matrix

$$\Pi = (I_2, B) = \begin{pmatrix} 1 & 0 & p_1 + ip_2 & r_1 + ir_2 \\ 0 & 1 & q_1 + iq_2 & s_1 + is_2 \end{pmatrix} .$$

We have

$$B_1 = \mathrm{Re}\ B = \begin{pmatrix} p_1 & r_1 \\ q_1 & s_1 \end{pmatrix} ,$$

$$B_2 = \mathrm{Im}\ B = \begin{pmatrix} p_2 & r_2 \\ q_2 & s_2 \end{pmatrix}$$

and we can still choose $B$ such that $D = \det B_2 > 0$. Consider the vector space $\mathbb{C}^2$ to be $\mathbb{R}^4$ with the complex structure given by the matrix

$$J = \begin{pmatrix} 0 & -I_2 \\ I_2 & 0 \end{pmatrix}$$

and take on $\mathbb{R}^4$ the complex structure given by the matrix

$$J_B = \begin{pmatrix} -B_1 B_2^{-1} & -B_2 - B_1 B_2^{-1} B_1 \\ B_2^{-1} & B_2^{-1} B_1 \end{pmatrix} .$$

Let $f : \mathbb{R}^4 \to \mathbb{C}^2$ be the map given by the matrix

$$F = \begin{pmatrix} I_2 & B_1 \\ 0 & B_2 \end{pmatrix}.$$

Then $F J_B = J F$ and, since $f(\mathbb{Z}^4) = \Gamma$, the map $f$ induces an analytic isomorphism between the topological torus $(\mathbb{R}/\mathbb{Z})^4$ with the complex structure given by the matrix $J_B$ and the complex torus $X = \mathbb{C}^2/\Gamma$. By the Appell-Humbert Theorem we get (see [Sl]):

$$NS(X) \cong \left\{ A = \begin{pmatrix} A_1 & A_2 \\ -A_2^t & A_3 \end{pmatrix} \in \mathcal{M}_4(\mathbb{Z}) \ \Big| \ \begin{matrix} A \text{ skew-symmetric and} \\ B^t A_1 B + A_2^t B - B^t A_2 + A_3 = 0 \end{matrix} \right\}.$$

The condition

$$B^t A_1 B + A_2^t B - B^t A_2 + A_3 = 0$$

expresses the fact that $A$ is the imaginary part $\operatorname{Im} H$ of a hermitian form $H$ on $\mathbb{C}^2$. The matrix of this hermitian form in the canonical basis of $\mathbb{C}^2$ is the following

$$H_A = (A_1 B_1 - A_2) B_2^{-1} + i A_1 .$$

Choosing, for instance $B_1 = 0$ and

$$B_2 = \begin{pmatrix} 1 & -\sqrt{2} \\ \sqrt{2} & 1 \end{pmatrix},$$

one can easily see that the corresponding 2-torus $X$ has algebraic dimension $a(X) = 0$. For $B_1 = 0$ and

$$B_2 = \begin{pmatrix} 1 & \sqrt{2} \\ 0 & 1 \end{pmatrix}$$

we get that the corresponding 2-torus $X$ has algebraic dimension $a(X) = 1$. Finally, taking $B_1 = 0$ and

$$B_2 = \begin{pmatrix} 1 & 0 \\ 0 & 1 \end{pmatrix}$$

we get a 2-torus $X$ with algebraic dimension $a(X) = 2$, i.e. an algebraic 2-torus. It is easy to see that for any 2-torus $X$ we have $b_1(X) = 4$ and $\mathcal{K}_X \cong \mathcal{O}_X$ (i.e. $\mathcal{K}_X$ is trivial).

(7) A *K3-surface* is a surface $X$ with $q(X) = \dim H^1(X, \mathcal{O}_X) = 0$ and $\mathcal{K}_X \cong \mathcal{O}_X$. They are all simply-connected and $H^2(X, \mathbb{Z})$ is torsion-free of rank 22. By standard arguments we get the following result (see [B-P-V], p. 241):

**Proposition 2.30** *Let $X$ be a K3-surface. Then:*

(1) *The map $c_1 : Pic(X) \to H^2(X, \mathbb{Z})$ is injective, hence maps $Pic(X)$ isomorphically onto the Neron-Severi group. In particular, an effective divisor is never homologous to zero.*

(2) *$h^0(X, T_X) = h^2(X, T_X) = 0$ , $h^1(X, T_X) = 20$, where $T_X$ is the tangent bundle of $X$.*

It was a hard job to prove that every K3-surface is Kählerian (see [Si], [B-P-V], p. 269). As examples of K3-surfaces we shall present here the so-called Kummer surfaces. Let $T$ be a 2-torus on which a base point 0 has been chosen. The involution $\tau : T \to T$, defined by $\tau(x) = -x$ has exactly sixteen fixed points, namely the points of order 2 on $T$. The quotient $T/G$, where $G$ is the group $\{1, \tau\}$, is a surface with sixteen singular points ( a 2-dimensional complex space...), all of them ordinary double points. Resolving the double points we obtain a smooth surface $X$, which is called the *Kummer surface* of $T$ and it is denoted by Km($T$). Let $g : X \to T/G$ be the projection. If we blow-up the sixteen fixed points of $\tau$ we get a surface $\tilde{T}$ such that the involution $\tau$ of $T$ can be lifted to an involution $\tilde{\tau}$ of $\tilde{T}$, which leaves the sixteen exceptional curves point-wise invariant but has otherwise no fix-points. The quotient of $\tilde{T}$ by the group $\{1, \tilde{\tau}\}$ is isomorphic, in fact, to $X = $ Km($T$) and we get the commutative diagram

$$
\begin{array}{ccc}
\tilde{T} & \longrightarrow & X = \tilde{T}/\{1, \tilde{\tau}\} \\
f \downarrow & & \downarrow g \\
T & \longrightarrow & T/\{1, \tau\}
\end{array}
$$

It is not hard to see that $q(X) = 0$ and $\mathcal{K}_X \cong \mathcal{O}_X$, i.e. $X = $ Km($T$) is a K3-surface. Of course, for the algebraic dimension $a($Km$(T)) = a(T)$, we get all three possible values $0, 1$ and $2$.

**(4)** An *Enriques surface* is a surface $X$ with $q(X) = 0$ ( or equivalently $b_1(X) = 0$) for which $\mathcal{K}_X^{\otimes 2} \cong \mathcal{O}_X$, but $\mathcal{K}_X \neq \mathcal{O}_X$. Now we shall recall the notion of cyclic covering. Let $Y$ be a connected complex manifold and $B$ a divisor on $Y$ which is either effective or zero. Suppose we have a line bundle $\mathcal{L}$ on $Y$ such that $\mathcal{O}_Y(B) \cong \mathcal{L}^{\otimes n}$, and a section $s \in H^0(Y, \mathcal{O}_Y(B))$ vanishing exactly along $B$ (if $B = 0$, we take for $s$ the constant function 1). Let $L$ be the total space of $\mathcal{L}$ and let $p : L \to Y$ be the bundle projection. If $t \in H^0(L, p^*\mathcal{L})$ is the tautological section, then the zero divisor of $p^*s - t^n$ defines an analytic subspace $X$ in $L$. If $B \neq 0$ and it is reduced, then $X$ is an irreducible normal analytic subspace of $L$ and $\pi = p|_X$ makes $X$ as an $n$-fold ramified covering of $Y$ with branch-locus $B$. We call $X \xrightarrow{\pi} Y$ the *$n$-cyclic covering* of $Y$ *branched* along $B$, *determined* by $\mathcal{L}$. If $B = 0$ we must take $n$ minimal (i.e. $\mathcal{L}$ is exactly of order $n$ in Pic$(Y)$) in order to obtain a connected manifold $X$. In this case $X \xrightarrow{\pi} Y$ is called the *$n$-cyclic unramified covering* of $Y$ determined by the torsion bundle $\mathcal{L}$. If Pic$(Y)$ has no torsion, then $B$ uniquely determines $\mathcal{L}$ and we may speak of the $n$-cyclic covering of $Y$, branched along $B$. It is clear that $X$ has at most singularities over singular points of $B$. In particular, if $B$ is reduced and smooth, then also $X$ is smooth. We have (see [B-P-V], p. 42):

**Lemma 2.31** *Let $\pi : X \to Y$ be the $n$-cyclic covering of $Y$ branched along a smooth divisor $B$ and determined by $\mathcal{L}$, where $\mathcal{L}^{\otimes n} = \mathcal{O}_Y(B)$. Let $B_1$ be the reduced divisor $\pi^{-1}(B)$ on $X$. Then:*

*(1) $\mathcal{O}_X(B_1) = \pi^*\mathcal{L}$;*

*(2) $\pi^* B = n B_1$ (in particular, $n$ is the branching order along $B_1$);*

*(3) $\mathcal{K}_X \cong \pi^*(\mathcal{K}_Y \otimes \mathcal{L}^{n-1})$;*

*(4) $\pi_* \mathcal{O}_X \cong \oplus_{j=0}^{n-1} \mathcal{L}^{-j}$.*

Now, if $B$ is a reduced divisor on the compact complex surface $Y$, such that $\mathcal{O}_Y(B) \cong \mathcal{L}^{\otimes 2}$ for some $\mathcal{L} \in \mathrm{Pic}(Y)$, then it determines a double covering $\pi : X \to Y$ which is ramified exactly over $B$. The surface $X$ is normal, and if $B$ has a simple singularity at $y_1 \in Y$, then $X$ has a rational singularity of the same type at $\pi^{-1}(y_1)$. If $\sigma : \tilde{X} \to X$ is the canonical resolution of singularities, and $p = \pi \circ \sigma$, we have

$$p_* \mathcal{O}_{\tilde{X}} = \mathcal{O}_Y \oplus \mathcal{L}^{-1} \ , \ R^i p_*(\mathcal{O}_{\tilde{X}}) = 0 \quad \text{for } i \geq 1.$$

If moreover $B$ has at most simple singularities, then

$$\mathcal{K}_{\tilde{X}} \cong p^*(\mathcal{K}_Y \otimes \mathcal{L}) \ .$$

(see [B-P-V], p. 182). If we take for $Y$ the quadric $\mathbb{P}^1 \times \mathbb{P}^1$ and for $B$ any curve of bidegree $(4,4)$ with at most simple singularities we find that the resulting surface $\tilde{X}$ of the above construction satisfies: $\mathcal{K}_{\tilde{X}} \cong \mathcal{O}_{\tilde{X}}$ and $q(\tilde{X}) = 0$, i.e. $\tilde{X}$ is a K3-surface (thus we have a new example of a K3-surface). On $\mathbb{P}^1 \times \mathbb{P}^1$ we have the involution $\tau$ given by $\tau(x_0 : x_1, y_0 : y_1) = (x_0 : -x_1, y_0 : -y_1)$, which has four isolated fix-points $p_i$. One can prove that we have a lot of $\tau$-invariant curves $B$ of bidegree $(4,4)$ not passing through any of the points $p_i$ and having at most simple singularities. Then we can see that $\tau$ lifts to a fixed-point-free involution $\tilde{\tau}$ of $\tilde{X}$. If $\pi_1 : \tilde{X} \to Y$ is the quotient map, then $Y$ is a compact surface with the following properties: $\mathcal{K}_Y \neq \mathcal{O}_Y$, $\mathcal{K}_Y^{\otimes 2} = \mathcal{O}_Y$ and $q(Y) = 0$ (see [B-P-V], p. 184). It follows that $Y$ is an Enriques surface.

An Enriques surface is projective. Since $p_g(Y) = 0$ it follows from the exponential cohomology sequence that for every $c \in H^2(Y, \mathbb{Z})$ there is a line bundle $\mathcal{L}$ on $Y$ with $c_1(\mathcal{L}) = c$. Standard arguments show that $b^+(Y) = 1$ ($b_1(Y) = 0$), hence there is a holomorphic line bundle $\mathcal{L}$ on $Y$ with $c_1^2(\mathcal{L}) > 0$. By Kodaira criterion the Enriques surface $Y$ is projective. It is not hard to prove the following result (see [B-P-V], p. 270):

**Proposition 2.32** *Let $Y$ be an Enriques surface. Then:*

*(1) The fundamental group of $Y$ is $\mathbb{Z}_2$ and the universal covering $X$ of $Y$ is a K3-surface;*

*(2) The map $\mathrm{Pic}(Y) \overset{c_1}{\to} H^2(Y, \mathbb{Z}) \cong \mathbb{Z}^{10} \oplus \mathbb{Z}_2$ is an isomorphism;*

*(3) The intersection form on $H^2(X, \mathbb{Z})/\mathrm{Tors}$ is isometric to the even unimodular lattice $(-E_8) \oplus H$;*

*(4) $h^0(Y, T_Y) = h^2(Y, T_Y) = 0$ , $h^1(Y, T_Y) = 10$.*

**(6)** A *primary Kodaira surface* is a surface with $b_1(X) = 3$ admitting a holomorphic locally trivial fibration over an elliptic curve with an elliptic curve as typical fibre. A *secondary Kodaira surface* is a surface, other than a primary Kodaira surface, admitting a primary Kodaira surface as unramified covering. They are elliptic fibre spaces over rational curves, with first Betti number equal to 1.

Consider, more generally, elliptic fibre bundles over curves (see [B-P-V], p. 143). If $E$ is an elliptic curve, we denote by $A(E)$ the group of its biholomorphic automorphisms. After fixing an origin $0 \in E$ this group can be described in the following way; $E$, acting on itself by translations, forms a normal subgroup of $A(E)$ and the quotient $A(E)/E$ can be identified with the group of automorphisms leaving $0$ fixed. So this quotient is the cyclic group $\mathbb{Z}_n$ of order 2, 4 or 6. Then $A(E)$ is the semi-direct product $E \times \mathbb{Z}_n$. The translation group $E$ is described by the universal covering sequence

$$0 \to \Gamma \to \mathbb{C} \to E \to 0 \,, \tag{2.3}$$

where $\Gamma$ is a lattice in $\mathbb{C}$. If $B$ is any smooth, compact, connected curve, then the holomorphic fibre bundles with typical fibre $E$ and base $B$ are classified by the cohomology set $H^1(B, \mathcal{A}_B)$ and there is an exact sequence of cohomology sets

$$H^1(B, \mathcal{E}_B) \to H^1(B, \mathcal{A}_B) \to H^1(B, \mathbb{Z}_n) \,,$$

where $\mathcal{A}_B$, resp. $\mathcal{E}_B$, is the sheaf of germs of local holomorphic maps from $B$ to $A(E)$, resp. $E$. We shall call a bundle $X \to B$ a *principal bundle* if its structure group can be reduced to $E$. To describe $H^1(B, \mathcal{E}_B)$ we use the exact cohomology sequence

$$H^1(B, \Gamma) \to H^1(B, \mathcal{O}_B) \to H^1(B, \mathcal{E}_B) \overset{c}{\to} H^2(B, \Gamma) \to 0 \,,$$

which is induced by (2.3). We have (see [B-P-V], p. 143):

**Lemma 2.33** *Let $X \to B$ be a holomorphic fibre bundle over the curve $B$ with typical fibre an elliptic curve $E$.*

*(1) The bundle $X \to B$ is principal if and only if $X$ admits an action of the group $E$ which on all the fibres $X_b$, $b \in B$, induces the translation group;*

*(2) Two principal $E$-bundles defined by cocycles $\xi = (\xi_{ij})$ and $\xi' = (\xi'_{ij})$ are isomorphic as $A(E)$-bundles if and only if there is some $z \in \mathbb{Z}_n$ such that $\xi' = z\xi$ in $H^1(B, \mathcal{E}_B)$;*

*(3) A principal $E$-bundle with class $\xi \in H^1(B, \mathcal{E}_B)$ can be defined by a locally constant cocycle if and only if $c(\xi) = 0$.*

It is not difficult to show that every principal bundle with typical fibre $E$ admits as unramified covering a holomorphic $\mathbb{C}^*$-bundle. Using this fact, Gysin sequence and the Künneth-formula one gets easily the following result:

**Proposition 2.34** *Let $X \to B$ be a principal $E$-bundle with class $\xi \in H^1(B, \mathcal{E}_B)$.*

*(1) If $c(\xi) = 0$, i.e. the principal bundle is topologically trivial, then $b_1(X) = b_1(B) + 2$ and $b_2(X) = 2b_1(B) + 2$;*

*(2) If $c(\xi) \neq 0$ (so the bundle is not topologically a product), then $b_1(X) = b_1(B) + 1$ and $b_2(X) = 2b_1(B)$.*

By a result of Höfer [Hf], every $E$-principal bundle (in fact, every torus-principal bundle) over a curve $B$ comes (in a unique way) from the following construction: there exist line bundles $\mathcal{L}_1, \cdots, \mathcal{L}_l$ over $B$ and elements $\gamma_1, \cdots, \gamma_l$ in $\Gamma$ such that the principal bundle is obtained by a *logarithmic transformation* applied to the trivial $E$-principal bundle $B \times E$; by choosing a sufficiently fine open covering $(U_i)$ of $B$ the transition functions of each $\mathcal{L}_k$ are expressed by a cocycle $(f_{ij}^k)$ and, by identifying $(z, t_i) \in U_i \times E$ with $(z, t_j) \in U_j \times E$ for all $z \in U_i \cap U_j$ if and only if

$$t_i = t_j + [\sum \frac{\gamma_k}{2\pi i} \log(f_{ij}^k)] .$$

If the base curve $B = \mathbb{P}^1$ any elliptic fibre bundle is either a product or a Hopf surface.

Let us assume that $B$ is an elliptic curve and that the principal bundle $\pi : X \to B$ is defined by a cocycle $\xi \in H^1(B, \mathcal{E}_B)$. If $c(\xi) = 0$, then it is easy to see that $X \to B$ is an elliptic 2-torus. The bundles defined by cocycles $\xi \in H^1(B, \mathcal{E}_B)$ with $c(\xi) \neq 0$ are precisely *all* the primary Kodaira surfaces. From the above proposition we get their topological invariants:

$$H^1(X, \mathbb{Z}) = \mathbb{Z}^3 , \ H^2(X, \mathbb{Z}) = \mathbb{Z}^4 \text{ or } \mathbb{Z}^4 \oplus \mathbb{Z}_m .$$

Notice that because of $b_1(X) = 3$, primary Kodaira surfaces are not kählerian. Since translations operate trivially on $H^1(E, \mathcal{O}_E)$ the line bundle $R^1\pi_*\mathcal{O}_X$ on $B$ is trivial. From the exact sequence

$$0 \to H^1(B, \mathcal{O}_B) \to H^1(X, \mathcal{O}_X) \to H^0(B, R^1\pi_*\mathcal{O}_X) \to 0$$

we find $h^1(X, \mathcal{O}_X) = 2$ and from

$$0 \to \pi^*(\omega_B) \to \Omega_X \to \omega_{X/B} \to 0$$

we deduce that $\mathcal{K}_X$ is trivial.

Sometimes a primary Kodaira surface admits a finite, freely-operating group of automorphisms. The smooth quotients thus obtained are called *secondary Kodaira surfaces*. Let $B$ be any elliptic curve and $p \in B$. We consider the line bundle $\mathcal{O}_B(p)$, and denote by $L$ the total space of the associated principal $\mathbb{C}^*$-bundle. Let $a \in \mathbb{C}^*$, $|a| \neq 1$, and let $g_a : L \to L$ be the automorphism obtained by multiplication with $a$ in each fibre. The quotient $X = L/ <g_a>$ is an elliptic fibre bundle over $B$. Since $c_1(\mathcal{O}_B(p)) \neq 0$ we get that $X$ is a primary Kodaira surface. Now, we take an involution $\tau : B \to B$ which has $p$ as a fix-point. Since $\tau^*(\mathcal{O}_B(p)) \cong \mathcal{O}_B(p)$, there exists a biholomorphic map $\alpha : L \to L$, covering $\tau$. Upon multiplication with a suitable automorphism of $L$ we obtain a biholomorphic map $\beta : L \to L$, covering $\tau$, with $\beta^2 = id_L$. Now we can take $\rho : X \to X$, $\rho = g_{\sqrt{a}} \circ \beta$, which is a fix-point-free involution. The quotient $Y = X/ <\rho>$ will be a secondary Kodaira surface. For a secondary Kodaira surface $Y$ we have: $b_1(Y) = 1$, $b_2(Y) = 0$, $q(Y) = 1$, $p_g(Y) = 0$.

(5) Take non-principal elliptic fibre bundles over an elliptic curve $B$. Then the bundle $\pi : X \to B$ is given by a class $\xi \in H^1(B, \mathcal{A}_B)$ which has a non-trivial image $\zeta \in H^1(B, \mathbb{Z}_n)$. It is not very difficult to prove that there is an elliptic curve $C$ such that $B = C/G$, where $G$ is a finite subgroup of the translation group of $C$, and that $X$ is the quotient $(E \times C)/G$, with $G$ acting on $C$ by translations and on $E$ by some representation $G \to A(E)$, which has its image not in the group of translations only. Such a surface is called *hyperelliptic*. The classification of these hyperelliptic surfaces contains seven subclasses precisely described (see [B-P-V], p. 148). Since the covering $E \times C$ is projective, then every hyperelliptic surface is projective (i.e. $a(X) = 2$). Its invariants are $h^{1,0} = 1$, $h^{2,0} = 0$, $h^{1,1} = 2$ and $\mathcal{K}_X$ is a torsion bundle of order 2, 3, 4 or 6.

(9) A *properly elliptic surface* is an elliptic surface $X$ with $\mathrm{kod}(X) = 1$. A very simple example is provided by the product of two curves, one elliptic and the other of genus $\geq 2$. Of course, the surface in this example has the algebraic dimension 2. In order to give some other examples of properly elliptic surfaces, consider again principal elliptic fibre bundles $X$ over a curve $B$ of genus $\geq 2$, given by $\xi \in H^1(B, \mathcal{E}_B)$ with $c(\xi) \neq 0$. It follows by Proposition 2.34 that $b_1(X)$ is odd, so $X$ is non-kählerian, hence $a(X) = 1$. Clearly, we have $\mathrm{kod}(X) = 1$.

(10) A *surface of general type* is a surface with $\mathrm{kod}(X) = 2$. Examples of such surfaces are complete intersections of sufficiently high degree, products of curves of genus $\geq 2$ and "practically all" ramified double coverings of $\mathbb{P}^2$. These surfaces are general in the same sense as are curves of genus $\geq 2$. Since always $a(X) \geq \mathrm{kod}(X)$, every surface of general type is algebraic. For more about surfaces of general type, see [B-P-V, Ct, Gi2, H-V, Se, Vv].

## 2.4 Intersection form and Neron-Severi group

In this section we shall precise the result from Corollary 2.9, i.e. the semi-negativity of the intersection form on the Neron-Severi group $\mathrm{NS}(X)$ of a nonalgebraic surface $X$. Firstly, we recall two important results. The inclusion of sheaves $i : \mathbb{Z}_X \to \mathbb{C}_X$ induces a homomorphism

$$i^* : H^2(X, \mathbb{Z}) \longrightarrow H^2(X, \mathbb{C}) .$$

We have (see, for instance [B-P-V], p. 119):

**Theorem 2.35 (Lefschetz Theorem on (1,1)-classes)** *Let $X$ be a compact complex surface. Then the image of the Picard group $\mathrm{Pic}(X)$ in $H^2(X, \mathbb{C})$ (i.e. $i^*(\mathrm{NS}(X))$) is $H^{1,1}(X) \cap i^*(H^2(X, \mathbb{Z}))$. In other words: an element of $H^2(X, \mathbb{C})$ is in the image of $\mathrm{Pic}(X)$ if and only if it is "integral" and can be represented by a real closed $(1,1)$-form.*

Let us denote by $H^{1,1}_{\mathbb{R}}(X) := H^{1,1}(X) \cap H^2(X, \mathbb{R})$. We have:

**Theorem 2.36 (Signature Theorem)** *Let $X$ be a compact complex surface. Then the cup-product form on $H^2(X, \mathbb{R})$, restricted to $H^{1,1}_{\mathbb{R}}(X)$, is non-degenerate of type $(1, h^{1,1} - 1)$ if $b_1(X)$ is even and of type $(0, h^{1,1})$ if $b_1(X)$ is odd.*

For nonalgebraic surfaces with algebraic dimension $a(X) = 0$ we can prove (see [B-F1, B-F3]):

**Theorem 2.37** *Let $X$ be a compact complex surface with $a(X) = 0$. Then the quadratic intersection form on the Neron-Severi group $NS(X)$ is negative-definite modulo torsion.*

*Proof.* Let $\tilde{X} \to X$ be the blow-up of $X$ in a point. Then we have (see Theorem 2.20)

$$\text{Pic}(\tilde{X}) \cong \text{Pic}(X) \oplus \mathbb{Z}e \ ,$$

where $e^2 = -1$ and $e.x = 0$ for any $x \in \text{Pic}(X)$. It follows that

$$NS(\tilde{X}) \cong NS(X) \oplus \mathbb{Z}e'$$

and the sum is orthogonal. Therefore it suffices to prove the statement for minimal models.

Let $X$ be a minimal model. If the Kodaira dimension $\text{kod}(X) = -\infty$, then we have $b_1(X) = 1$ and, by the Signature Theorem, it follows that the quadratic intersection form on $H^{1,1}_{\mathbb{R}}(X)$ is negative-definite. Then its restriction to the subgroup $NS(X) \subset H^2(X, \mathbb{Z})$ is negative-definite modulo torsion. If the Kodaira dimension $\text{kod}(X) = 0$, then $X$ is a K3-surface or a 2-torus. If $X$ is a K3-surface with $a(X) = 0$ then it is known that the quadratic intersection form is negative-definite (see, for instance [Be]). Here there is a shorter proof. Let $L \in \text{Pic}(X)$ such that $c_1^2(L) = 0$. From Riemann-Roch Theorem we get

$$h^0(X, L) + h^0(X, L^*) \geq 2 \ .$$

By Proposition 2.15 we have $h^0(X, M) \leq 1$ for every $M \in \text{Pic}(X)$, hence $h^0(X, L) = h^0(X, L^*) = 1$. It follows that $L \cong \mathcal{O}_X$, i.e. $c_1(L) = 0$. If $X$ is a 2-torus we have to recall some facts. Any element $A \in NS(X)$ is the first Chern class of a line bundle $L \in \text{Pic}(X)$ $(A = c_1(L))$. If we identify the group $H^2(X, \mathbb{Z})$ with $\text{Alt}^2_{\mathbb{Z}}(\Gamma, \mathbb{Z})$ (see the notations in Section 2.3), then the cup-product on $H^2(X, \mathbb{Z})$ is the exterior product of 2-forms (see Mumford [Mm4], p. 17). The intersection form on the group $NS(X)$ (with the notations of Section 2.3) is given by the formula

$$c_1(L).c_1(L') = \alpha\delta' + \alpha'\delta - \beta\gamma' - \beta'\gamma - \theta\tau' - \theta'\tau \ ,$$

where $c_1(L) = A$, $c_1(L') = A'$,

$$A = \begin{pmatrix} A_1 & A_2 \\ -A_2^t & A_3 \end{pmatrix}; \ A_1 = \begin{pmatrix} 0 & \theta \\ -\theta & 0 \end{pmatrix}, \ A_2 = \begin{pmatrix} \alpha & \beta \\ \gamma & \delta \end{pmatrix}, \ A_3 = \begin{pmatrix} 0 & \tau \\ -\tau & 0 \end{pmatrix},$$

with $A_1, A_2, A_3 \in M_2(\mathbb{Z})$ and similar formula for $A'$. For the quadratic intersection form we get

$$c_1^2(L) = 2(\alpha\delta - \beta\gamma - \theta\tau) \ .$$

By direct computation one obtains the formula (see [B-F1]):

$$c_1^2(L) = 2D.\det(H_A) \ ,$$

where $H_A$ is the hermitian matrix corresponding to the Chern class $A = c_1(L)$ (see Section 2.3).

Let us suppose that there exists $c_1(L) = A \in \text{NS}(X)$, $c_1(L) \neq 0$, such that $c_1^2(L) = 0$. By the previous formula we get $\det(H_A) = 0$ $(D > 0)$. The hermitian matrix $H_A$ is unitarily similar to a diagonal matrix

$$\begin{pmatrix} \lambda_1 & 0 \\ 0 & \lambda_2 \end{pmatrix},$$

where $\lambda_1, \lambda_2 \in \mathbb{R}$ are the eigenvalues of the matrix $H_A$. Since $\lambda_1 \lambda_2 = \det(H_A) = 0$, it follows (say) $\lambda_1 = 0$. Because $c_1(L) = A \neq 0$, hence $H_A \neq 0$ and $\lambda_2 \neq 0$. By changing (if necessary) $A$ with $-A$ (and $H_A$ with $-H_A$) we can suppose that $\lambda_2 > 0$. Then the hermitian matrix $H_A$ is positive semi-definite and, by [We], it follows that $a(X) \geq 1$, contradiction. Since $\text{kod}(X) \geq 1$ implies $a(X) \geq 1$ the proof is over.

Let now $X$ be a compact complex surface with $a(X) = 1$. One knows that there exists an unique elliptic fibration $f : X \to S$ (see Theorem 2.13 and Proposition 2.14). Let $C$ be a general fibre of $f$; then we have $c_1^2(\mathcal{O}_X(C)) = C^2 = 0$. For any $L \in \text{Pic}(X)$ with $c_1^2(L) = 0$, the Chern class $c_1(L)$ is orthogonal to the group $\text{NS}(X)$ (Corollary 2.10), hence $c = c_1(\mathcal{O}_X(C))$ is orthogonal to $\text{NS}(X)$. We have in this case (see [B-F3]):

**Theorem 2.38** *Let $X$ be a compact complex surface with $a(X) = 1$. Then we have an orthogonal sum $\text{NS}(X)/\text{Tors}\,\text{NS}(X) = I \oplus N$ such that $I$ is an isotropic subgroup of rank $\leq 1$ and the quadratic intersection form is negative-definite on $N$. Moreover:*

*(1) If $b_1(X)$ is odd, then $I = 0$;*

*(2) If $X$ is Kähler (i.e. $b_1(X)$ is even), then $I$ is generated by a rational multiple of $c = c_1(\mathcal{O}_X(C))$;*

*(3) If $X$ has as minimal model a K3-surface or a 2-torus, then $I$ is generated by $c = c_1(\mathcal{O}_X(C))$.*

*Proof.* Let us denote by $J$ the lattice $\text{NS}(X)/\text{Tors}\,\text{NS}(X)$. Let $I = \text{Rad}\,J$ be the radical of $J$ and let $J = I \oplus N$ be a radical splitting (orthogonal sum). If $b_1(X)$ is even then by Signature Theorem it follows that the intersection form on $H_{\mathbb{R}}^{1,1}(X)$ is non-degenerate of type $(1, h^{1,1} - 1)$. Clearly, $H_{\mathbb{R}}^{1,1}(X)$ has the isotropy index 1, i.e. a maximal isotropic subspace has the dimension 1. Hence, the isotropy subgroup $I$ has rank $\leq 1$ and the quadratic intersection form is negative-definite on the factor $N$. If $b_1(X)$ is odd then by Signature Theorem it follows that the intersection form on $H_{\mathbb{R}}^{1,1}(X)$ is non-degenerate of type $(0, h^{1,1})$. Hence, we get $I = 0$ and the quadratic intersection form on $\text{NS}(X)$ is negative-definite modulo torsion ($c = c_1(\mathcal{O}_X(C))$ is a torsion element). These facts prove the first statement and (1).

If $X$ is Kähler (i.e. $b_1(X)$ even) then $c = c_1(\mathcal{O}_X(C)) \neq 0$ for the general fibre $C$ of the elliptic fibration $f : X \to S$. Since $c$ is not a torsion element of the group $\text{NS}(X)$ it follows (2). In order to prove (3) we can suppose that $f$ is relatively minimal (i.e. the fibres do not contain $(-1)$-curves). Since $\text{NS}(X)$ has no torsion for K3-surfaces and 2-tori, we get the orthogonal sum $\text{NS}(X) = I \oplus N$. We have to prove that $I$ is generated by the element $c = c_1(\mathcal{O}_X(C))$. Let $d \neq 0$ be an element of $I$ $(d^2 = 0)$. Let us suppose that $X$ is a K3-surface and let us take $L \in \text{Pic}(X)$ such that $d = c_1(L)$.

By Riemann-Roch Theorem we have $h^0(X, L) + h^0(X, L^*) \geq 2$; it follows $d$ or $-d$ effective. Suppose $d = c_1(\mathcal{O}_X(D))$, where $D$ is an effective divisor. Since all curves on $X$ are contained in the fibres of $f$ (cf. Proposition 2.14) we get $D = D_1 + \cdots + D_n$, with every effective divisor $D_i$, $i = 1, \cdots, n$, contained in different fibres of $f$. Clearly, $D_i.D_j = 0$ for $i \neq j$, hence

$$D^2 = D_1^2 + \cdots + D_n^2 = 0 .$$

From $D_i^2 \leq 0$, $i = 1, \cdots, n$ (Corollary 2.9) it follows $D_i^2 = 0$, $i = 1, \cdots, n$. Thus we can suppose $D$ contained in a fibre. By Zariski Lemma (see [B-P-V], p. 90), it follows $pD = qX_s$, with $p, q \in \mathbb{Z}$, $p \neq 0$ and $X_s$ a fibre of $f$ ($s \in S$). One knows that $f$ has no multiple fibres (see [B-P-V], p. 195), hence $p = 1$ and $D = qX_s$. Since $c_1(\mathcal{O}_X(X_s)) = c_1(\mathcal{O}_X(C)) = c$, we get that $c$ generates the isotropic subgroup $I$. Let now $X$ be a 2-torus. One knows that $f$ has no singular fibres and, topologically, $X$ is the product $C \times S$. By Künneth formula we have

$$H^2(X, \mathbb{Z}) \cong H^2(C, \mathbb{Z}) \oplus (H^1(C, \mathbb{Z}) \otimes H^1(S, \mathbb{Z})) \oplus H^2(S, \mathbb{Z}) ,$$

where the subgroup $H^2(C, \mathbb{Z})$ is generated by the Chern class $c = c_1(\mathcal{O}_C(C))$. It follows that the group $H^2(X, \mathbb{Z})/c\mathbb{Z}$ has no torsion, hence the subgroup $I/c\mathbb{Z}$ has no torsion. Then $I$ is generated by $c$, hence (3).

# 3. Line bundles over surfaces

In this chapter we present solutions to the existence and classification problems for holomorphic structures in topological line bundles. Then we compute the Neron-Severi group for tori and some elliptic surfaces, namely elliptic bundles and quasi-bundles over curves. Finally, we present the computation of the Picard group for tori (Appell-Humbert Theorem) and we compute the Picard group for primary Kodaira surfaces.

## 3.1 Holomorphic structures in line bundles

Let $X$ be a compact complex manifold and let $\mathcal{O}_X$ denote its structural sheaf. Recall that there is an isomorphism

$$\mathrm{Vect}^1_{top}(X) \cong H^1(X, \mathcal{C}^*) \xrightarrow{\delta} H^2(X, \mathbb{Z})$$

between the group of classes of topological (smooth) line bundles over $X$ and the second integer cohomology group of $X$, where $\delta(L) = c_1(L)$ for any $L \in \mathrm{Vect}^1_{top}(X)$. Recall also that the exponential sequence

$$0 \to \mathbb{Z} \to \mathcal{O}_X \xrightarrow{\exp} \mathcal{O}_X^* \to 0$$

gives rise to the exact cohomology sequence

$$\to H^1(X, \mathbb{Z}) \to H^1(X, \mathcal{O}_X) \to H^1(X, \mathcal{O}_X^*) \xrightarrow{c_1} H^2(X, \mathbb{Z}) \to H^2(X, \mathcal{O}_X) \to,$$

and that the group $H^1(X, \mathcal{O}_X^*)$ is isomorphic to the Picard group $\mathrm{Pic}(X) = \mathrm{Vect}^1_{hol}(X)$. The image of $c_1$ is called the Neron-Severi group of $X$ and is denoted by $\mathrm{NS}(X)$. Then:

*Remark.* The answer to the existence of holomorphic structures in a given topological line bundle $L$ is very simple: we have to impose the condition that the Chern class $c_1(L) \in \mathrm{NS}(X)$.

We have the exact sequence

$$0 \to \mathrm{Pic}_0(X) \to \mathrm{Pic}(X) \xrightarrow{c_1} \mathrm{NS}(X) \to 0 \,,$$

where

$$\mathrm{Pic}_0(X) = \mathrm{Ker}(H^1(X, \mathcal{O}_X^*) \xrightarrow{c_1} H^2(X, \mathbb{Z})) \,.$$

**Proposition 3.1** *Let $X$ be a compact complex manifold. Then the group $\mathrm{Pic}_0(X)$ has always a structure of complex Lie group.*

*Proof.* It is enough to prove that the image of $H^1(X, \mathbb{Z})$ in $H^1(X, \mathcal{O}_X)$ is closed (since $\mathrm{Pic}_0(X) \cong H^1(X, \mathcal{O}_X)/\mathrm{Im}(H^1(X, \mathbb{Z}) \to H^1(X, \mathcal{O}_X)))$. But $H^1(X, \mathbb{Z})$ is torsion free and it is obviously closed in $H^1(X, \mathbb{R})$ (universal coefficients formula, $H^1(X, \mathbb{R}) \cong H^1(X, \mathbb{Z}) \otimes_{\mathbb{Z}} \mathbb{R}$), so it is enough to prove that $H^1(X, \mathbb{R})$ is mapped injectively in $H^1(X, \mathcal{O}_X)$ under the natural homomorphism

$$H^1(X, \mathbb{R}) \xrightarrow{j} H^1(X, \mathcal{O}_X) \,.$$

We shall use Čech cohomology (pointed out by O. Forster). Let $\mathcal{U} = (U_k)$ be an open covering of $X$ and let $r = \mathrm{cls.}(r_{kl}) \in H^1(\mathcal{U}, \mathbb{R})$ such that $j(r) = 0$. Then, there are holomorphic functions $f_k \in \mathcal{O}_X(U_k)$ such that $r_{kl} = f_l - f_k$ on $U_k \cap U_l$, for any $k, l$. Writing $f_k = u_k + iv_k$, we get $v_l = v_k$ on $U_k \cap U_l$, since $r_{kl}$ are real. It follows that there exists a global pluriharmonic function $v$ such that $v|_{U_k} = v_k$. Then, there exists a global holomorphic function $f$ with $\mathrm{Im}\, f = v$. Since $X$ is compact, $f$ must be a constant. We get the equalities $r_{kl} = u_l - u_k$ with $u_k \in \mathbb{R}$, i.e. $r = \mathrm{cls.}(r_{kl}) = 0$.

*Remark.* For any fixed $\eta \in \mathrm{NS}(X)$ define

$$\mathrm{Pic}^\eta(X) := c_1^{-1}(\eta) = \{L \in \mathrm{Pic}(X) \mid c_1(L) = \eta \} \,.$$

Obviously, we have a bijection $\mathrm{Pic}_0(X) \simeq \mathrm{Pic}^\eta(X)$, $M \mapsto L \otimes M$, where $M \in \mathrm{Pic}_0(X)$ and $L \in \mathrm{Pic}^\eta(X)$ is fixed. Observe that $\mathrm{Pic}(X)$ is a disjoint union

$$\mathrm{Pic}(X) = \coprod_{\eta \in NS(X)} \mathrm{Pic}^\eta(X) \,.$$

Thus, by the above proposition, we answer affirmatively to the Problem II of classification of holomorphic structures in line bundles.

Now, we want to give an answer to the more precise classification problem, namely, to Problem III.

Let $X$ be as before a compact complex manifold. Let $S$ be a complex space and let $p : S \times X \to S$ be the first projection.

**Definition 3.2 (Grothendieck)** If $\mathcal{A}n$ denotes the category of complex spaces, then

$$\underline{Pic}_X : \mathcal{A}n \longrightarrow Sets$$

is the (contravariant) *Picard functor*, defined by

$$\underline{Pic}_X(S) := H^0(S, R^1 p_*(\mathcal{O}^*_{S \times X})) \,.$$

We shall study the representability of this functor. Firstly, we remark that the functor $\underline{Pic}_X$ takes values in the category of abelian groups so, if it is representable by $P_X \in \mathcal{A}n$, then $P_X$ is an abelian group in the category $\mathcal{A}n$ of complex spaces (i.e. an *abelian analytic group*). Secondly, we observe that the above definition agree with the definition of families of holomorphic structures parametrised by a complex space $S$. Indeed, since the sheaf $R^1 p_*(\mathcal{O}^*_{S \times X})$ is the associated sheaf of the presheaf

$$U \mapsto H^1(U \times X, \mathcal{O}^*_{S \times X}) = \mathrm{Pic}(U \times X) \,, \; U \subset S \text{ open}$$

then, an element $u$ of $\underline{Pic}_X(S)$ is given by an open covering $(U_i)$ of $S$ and holomorphic line bundles $\mathcal{L}_i$ over $U_i \times X$, such that for any $i, j$

$$\mathcal{L}_i|_{(U_i \cap U_j) \times X} \text{ is locally isomorphic to } \mathcal{L}_j|_{(U_i \cap U_j) \times X}.$$

Now, we have to glue together the line bundles $\mathcal{L}_i$ in order to get a family of line bundles.

Consider the Leray spectral sequence (see [Go], Chapter II, 4.17) for $p : S \times X \to S$ and the sheaf $\mathcal{O}^*_{S \times X}$

$$E^{pq}_2 = H^p(S, R^q p_*(\mathcal{O}^*_{S \times X})) \Longrightarrow H^{p+q}(S \times X, \mathcal{O}^*_{S \times X})$$

and take the exact sequence of small terms of this spectral sequence

$$0 \to H^1(S, p_*(\mathcal{O}^*_{S \times X})) \to H^1(S \times X, \mathcal{O}^*_{S \times X}) \to H^0(S, R^1 p_*(\mathcal{O}^*_{S \times X})) \to$$

$$\to H^2(S, p_*(\mathcal{O}^*_{S \times X})) \to H^2(S \times X, \mathcal{O}^*_{S \times X}) \,.$$

Since $p_*(\mathcal{O}_{S \times X}) \cong \mathcal{O}_S$, hence $p_*(\mathcal{O}^*_{S \times X}) \cong \mathcal{O}^*_S$ and the above exact sequence has the form

$$0 \to \mathrm{Pic}(S) \to \mathrm{Pic}(S \times X) \to \underline{Pic}_X(S) \overset{\alpha}{\to} H^2(S, \mathcal{O}^*_S) \overset{p^*}{\to} H^2(S \times X, \mathcal{O}^*_{S \times X}).$$

Thus, the obstruction $\alpha(u)$ to lift an element $u \in \underline{Pic}_X(S)$ to a family $\mathcal{L} \in \mathrm{Pic}(S \times X)$ lives in $H^2(S, \mathcal{O}^*_S)$. From the exact sequence we have $p^*(\alpha(u)) = 0$. But $p : S \times X \to S$ has a (lot of) section(s) $g : S \to S \times X$ ($p \circ g = id_S$). It follows that $p^*$ is injective, hence $\alpha(u) = 0$. Finally, we get the exact sequence

$$0 \to \mathrm{Pic}(S) \to \mathrm{Pic}(S \times X) \to \underline{Pic}_X(S) \to 0 \,,$$

i.e. any element $u \in \underline{Pic}_X(S)$ is a class of equivalent families of holomorphic line bundles as in Definition 1.39.

We have (see Grothendieck [Gk4]):

**Theorem 3.3** *Let $X$ be a compact complex manifold. Then the Picard functor $\underline{Pic}_X$ is representable by an abelian analytic group $P_X$. Consider $P := \mathrm{Coker}(H^1(X, \mathbb{Z}) \to H^1(X, \mathcal{O}_X))$ and $Q := \mathrm{Ker}(H^2(X, \mathbb{Z}) \to H^2(X, \mathcal{O}_X))$ as abelian analytic groups ($P = Pic_0(X)$ is a connected complex Lie group by Proposition 3.1 and $Q$ is discrete). Then we have a canonical exact sequence of analytic groups*

$$0 \to P \to P_X \to Q \to 0 \,.$$

*Proof.* (cf. [Gk4]) For any complex space $S$ we have the exact sequence

$$0 \to \mathbb{Z}_{S \times X} \to \mathcal{O}_{S \times X} \overset{\exp}{\to} \mathcal{O}^*_{S \times X} \to 0 \,,$$

and we get the exact sequence of sheaves over $S$:

$$R^1 p_*(\mathbb{Z}_{S \times X}) \to R^1 p_*(\mathcal{O}_{S \times X}) \to R^1 p_*(\mathcal{O}^*_{S \times X}) \to R^2 p_*(\mathbb{Z}_{S \times X}) \to R^2 p_*(\mathcal{O}_{S \times X}),$$

where $p : S \times X \to S$ is the first projection. Obviously, we have isomorphisms

$$R^i p_*(\mathcal{O}_{S \times X}) \xrightarrow{\sim} \mathcal{O}_S \otimes_{\mathbb{C}} H^i(X, \mathcal{O}_X)$$

$$R^i p_*(\mathbb{Z}_{S \times X}) \xrightarrow{\sim} \mathbb{Z}_S \otimes_{\mathbb{Z}} H^i(X, \mathbb{Z}_X) \,.$$

Since by Proposition 3.1 the image of the canonical homomorphism

$$H^1(X, \mathbb{Z}) \to H^1(X, \mathcal{O}_X)$$

is closed, hence the cokernel

$$P := \mathrm{Coker}(H^1(X, \mathbb{Z}) \to H^1(X, \mathcal{O}_X))$$

is a connected analytic group. Let

$$Q := \mathrm{Ker}(H^2(X, \mathbb{Z}) \to H^2(X, \mathcal{O}_X))$$

be a discrete analytic group. Then we get an exact sequence:

$$0 \to \mathcal{O}_S \otimes_{\mathbb{C}} P \to R^1 p_*(\mathcal{O}^*_{S \times X}) \to \mathbb{Z}_S \otimes_{\mathbb{Z}} Q \to 0 \,, \qquad (3.1)$$

which is obviously functorial in $S$. Let

$$h_P \,, \ h_Q : \mathcal{A}n \longrightarrow Ab$$

be functors on the category of complex spaces $\mathcal{A}n$ with values in the category of abelian groups $Ab$, represented by $P$ and $Q$. Then we have

$$h_P(S) = \mathrm{Hom}(S, P) \cong H^0(S, \mathcal{O}_S \otimes_{\mathbb{C}} P) \,,$$

$$h_Q(S) = \mathrm{Hom}(S, Q) \cong H^0(S, \mathbb{Z}_S \otimes_{\mathbb{Z}} Q) \,.$$

From (3.1) we get an exact sequence of homomorphisms

$$0 \to h_P \to \underline{Pic}_X \to h_Q \,. \qquad (3.2)$$

To see that $\underline{Pic}_X$ is representable it suffices to prove it is representable relatively to $h_Q$ (see [Gk4, Gk5]), i.e. for any complex space $S$ and any element $\eta \in h_Q(S) \cong H^0(S, \mathbb{Z}_S \otimes_{\mathbb{Z}} Q)$, the functor $F$ on $\mathcal{A}n/S$, which associates to any $T$ over $S$ ($T \to S$) the set $F_\eta(T)$ of sections of $R^1 p'_*(\mathcal{O}^*_{T \times X})$ over $T$ ($p' : T \times X \to T$ is the first projection), whose images in $H^0(T, \mathbb{Z}_T \otimes_{\mathbb{Z}} Q)$ are the inverse images of $\eta$ by $T \to S$, is representable. Let $M$ be the subsheaf of sets of $R^1 p_*(\mathcal{O}^*_{S \times X})$ which are the inverse images of $\eta$ by the homomorphism

$$R^1 p_*(\mathcal{O}^*_{S \times X}) \to \mathbb{Z}_S \otimes_{\mathbb{Z}} Q.$$

Because of the exact sequence (3.1), it follows that $M$ is a homogeneous principal sheaf on $\mathcal{O}_S \otimes_{\mathbb{C}} P$, hence it defines an analytic homogeneous principal bundle $\mathcal{M}$ over $S$ with structural group $P$, which represents $F$. This proves the existence of $P_X$, which represents the Picard functor $\underline{Pic}_X$. Moreover, we get an exact sequence

$$0 \to P \to P_X \to Q \to 0 \,,$$

where $P_X$ is a homogeneous principal bundle over $Q$ with structural group $P$.

*Remark.* Of course, $P_X$ is nothing else as $\mathrm{Pic}(X) = H^1(X, \mathcal{O}^*_X)$, but we proved that it represents the Picard functor and thus, we have an universal line bundle (called *Poincaré bundle*) $\mathcal{P}_X$ on $P_X \times X$, etc.

For the case of an algebraic surface see, for example [Mm3] and, for the more general case of schemes see [A-K, Kl].

## 3.2 Picard group for tori

In this section we shall present the "explicit" computation of the Picard group of a complex torus (there are no differences between the case of a 2-torus and the general case) giving in this way a concrete answer to the question of *classifying all the holomorphic line bundles* over a torus. The result will be the classical Appell-Humbert Theorem (see also Section 2.3) and we shall give only the general ideas, following closely the presentation of Mumford [Mm4], Chapter I.

Let $X = \mathbb{C}^n/\Gamma$ be an $n$-dimensional complex torus, where $\Gamma$ is a lattice of real rank $2n$. Let $\pi : \mathbb{C}^n \to X$ be the natural projection. Since $\mathbb{C}^n$ is contractible we have

$$H^k(\mathbb{C}^n, \mathbb{Z}) = 0, \quad k \geq 1.$$

But we have also

$$H^k(\mathbb{C}^n, \mathcal{O}) = 0, \quad k \geq 1.$$

Then, using the exponential sequence we obtain $H^1(\mathbb{C}^n, \mathcal{O}^*) = 1$, i.e. every holomorphic line bundle over $\mathbb{C}^n$ is trivial. In fact, there is a more general result of Grauert [Gr1]; for a Stein manifold (or space) the topological classification and the holomorphic classification of the complex vector bundles is the same. More precisely: two holomorphic vector bundles which are topologically isomorphic are also analytically isomorphic and every topological vector bundle admits a holomorphic structure.

Let $L$ be a holomorphic line bundle over $X$. It follows that $\pi^*(L)$ is trivial over $\mathbb{C}^n$ and we choose an isomorphism

$$\chi : \pi^*(L) \overset{\sim}{\to} \mathbb{C}^n \times \mathbb{C} \,.$$

On $\pi^*(L)$ we have a linear (on fibres) action of $\Gamma$ such that $L$ is the quotient of $\pi^*(L)$ by this action. Through the isomorphism $\chi$ we get a linear action on $\mathbb{C}^n \times \mathbb{C}$ covering the action of $\Gamma$ on $\mathbb{C}^n$ such that the quotient is $L$. Let us denote by $H^*$ the multiplicative group $H^0(\mathbb{C}^n, \mathcal{O}^*)$ of non-vanishing holomorphic functions on $\mathbb{C}^n$. Then the action of $\Gamma$ on $\mathbb{C}^n \times \mathbb{C}$ is given by

$$\Phi_\gamma(z, t) = (z + \gamma, e_\gamma(z)t), \quad \gamma \in \Gamma \,,$$

where $e_\gamma \in H^*$. The condition that $\Phi_\gamma$ is a group action gives that $\gamma \mapsto e_\gamma$ is a 1-cocycle for the group $\Gamma$ with coefficients in $H^*$:

$$e_{\gamma+\gamma'}(z) = e_\gamma(z + \gamma')e_{\gamma'}(z), \quad \gamma, \gamma' \in \Gamma \,.$$

In fact, we get an isomorphism (see [Mm4], p. 14 or [Gk2]):

$$\Phi : H^1(\Gamma, H^*) \overset{\sim}{\to} H^1(X, \mathcal{O}_X^*) \,.$$

Now, our problem is to find the simplest kind of representing cocycles $(e_\gamma)$ for all cohomology classes in $H^1(X, \mathcal{O}_X^*)$.

If $H$ is the ring of holomorphic functions on $\mathbb{C}^n$, we have the exact sequence

$$0 \to \mathbb{Z} \to H \overset{\exp}{\to} H^* \to 0 \,,$$

and by [Mm4], p. 16, we get the commutative diagram:

$$H^1(\Gamma, H^*) \xrightarrow{\ \delta\ } H^2(\Gamma, \mathbb{Z})$$

$$\wr \downarrow \qquad\qquad \wr \downarrow$$

$$H^1(X, \mathcal{O}_X^*) \xrightarrow{\ c_1\ } H^2(X, \mathbb{Z})$$

By the above identifications it follows that the Chern class of $L$ is $\delta(\mathrm{cls}.(e_\gamma))$. If we write

$$e_\gamma(z) = e^{2\pi i f_\gamma(z)},$$

with $f_\gamma$ holomorphic in $\mathbb{C}^n$ then, by definition,

$$\delta(\mathrm{cls}.(e_\gamma)) \in H^2(\Gamma, \mathbb{Z})$$

is given by the 2-cocycle on $\Gamma$ with coefficients in $\mathbb{Z}$ given by

$$F(\gamma, \gamma') = f_{\gamma'}(z + \gamma) - f_{\gamma+\gamma'}(z) + f_\gamma(z) \in \mathbb{Z}. \tag{3.3}$$

Let $F : \Gamma \times \Gamma \to \mathbb{Z}$ be any map and let us denote by $AF : \Gamma \times \Gamma \to \mathbb{Z}$ the map

$$AF(\gamma, \gamma') = F(\gamma, \gamma') - F(\gamma', \gamma), \quad \gamma, \gamma' \in \Gamma.$$

Then we have:

**Lemma 3.4** *The function $A$ maps the group of 2-cocycles $Z^2(\Gamma, \mathbb{Z})$ in the group $Alt^2_{\mathbb{Z}}(\Gamma, \mathbb{Z})$ of alternating integer-valued 2-forms on $\Gamma$ and induces an isomorphism*

$$A : H^2(\Gamma, \mathbb{Z}) \xrightarrow{\sim} Alt^2_{\mathbb{Z}}(\Gamma, \mathbb{Z}) \cong \bigwedge^2 Hom(\Gamma, \mathbb{Z}).$$

*Moreover, we have*

$$A(\xi \cup \eta) = \xi \wedge \eta,$$

*for any $\xi, \eta \in Hom(\Gamma, \mathbb{Z}) \cong H^1(\Gamma, \mathbb{Z})$.*

For $F \in Z^2(\Gamma, \mathbb{Z})$ denote $E = AF$. Then, from the fact that $F$ is a cocycle we get immediately that $E$ is bilinear and alternating. Also, it is obvious that if $F$ is a coboundary, then $AF = 0$. Hence $A$ induces a homomorphism

$$H^2(\Gamma, \mathbb{Z}) \longrightarrow Alt^2_{\mathbb{Z}}(\Gamma, \mathbb{Z}) \cong \bigwedge^2 Hom(\Gamma, \mathbb{Z}).$$

Since we have an isomorphism of $H^*(\Gamma, \mathbb{Z})$ onto $H^*(X, \mathbb{Z})$, taking cup-products to cup-products (see [Mm4], p. 17), and since we know that $H^*(X, \mathbb{Z})$ is the exterior algebra on $H^1(X, \mathbb{Z})$, it follows that $H^*(\Gamma, \mathbb{Z})$ is the exterior algebra on $H^1(\Gamma, \mathbb{Z}) \cong Hom(\Gamma, \mathbb{Z})$. For any $\xi, \eta \in Hom(\Gamma, \mathbb{Z})$ we have that $\xi \cup \eta$ is given by the 2-cocycle $c(s, t) = \xi(s)\eta(t)$, so that

$$A(\xi \cup \eta)(s, t) = (\xi \wedge \eta)(s, t)$$

and the lemma follows.

We get the following

**Corollary 3.5** *The Chern class of the line bundle corresponding to the 1-cocycle* $(e_\gamma) \in Z^1(\Gamma, H^*)$ *is the alternating integer-valued 2-form given by*

$$E(\gamma, \gamma') = f_{\gamma'}(z + \gamma) + f_\gamma(z) - f_\gamma(z + \gamma') - f_{\gamma'}(z), \quad \forall z \in \mathbb{C}^n, \tag{3.4}$$

*where*

$$e_\gamma(z) = e^{2\pi i f_\gamma(z)}.$$

Since $E$ belongs to the Neron-Severi group, its image by $H^2(X, \mathbb{Z}) \to H^2(X, \mathcal{O}_X)$ must be zero. This map factorises naturally as

$$H^2(X, \mathbb{Z}) \xrightarrow{a} H^2(X, \mathbb{C}) \xrightarrow{j} H^2(X, \mathcal{O}_X).$$

Let

$$\text{Hom}_{\mathbb{R}}(\mathbb{C}^n, \mathbb{C}) = \text{Hom}_{\mathbb{C}}(\mathbb{C}^n, \mathbb{C}) \oplus \text{Hom}_{\mathbb{C}-anti}(\mathbb{C}^n, \mathbb{C}) = T \oplus \overline{T}.$$

Then we have the Hodge decomposition for the torus $X = \mathbb{C}^n/\Gamma$:

$$H^2(X, \mathbb{C}) \cong \bigwedge^2 (T \oplus \overline{T}) \cong \bigwedge^2 T \oplus (T \otimes \overline{T}) \oplus \bigwedge^2 \overline{T},$$

where $H^2(X, \mathcal{O}_X) \cong \bigwedge^2 \overline{T}$ and $j$ gives the natural projection. Then $a(E)$ is the real linear extension and we get easily that $j(E) = 0$ if and only if $E(x, y) = E(ix, iy)$ for all $x, y \in \mathbb{C}^n$. Thus, we have:

**Corollary 3.6** *If we extend the Chern class $E$ $\mathbb{R}$-linearly to a map $\mathbb{C}^n \times \mathbb{C}^n \to \mathbb{R}$, then $E$ satisfies the identity $E(ix, iy) = E(x, y)$ for all $x, y \in \mathbb{C}^n$.*

*Remark.* We shall use the following elementary fact: Let $V$ be a complex vector space. There is a $1 - 1$ correspondence between the Hermitian forms $H$ on $V$ and the real skew-symmetric forms $E$ on $V$ satisfying the identity $E(ix, iy) = E(x, y)$, which is given by

$$E(x, y) = \text{Im}(H(x, y)), \quad H(x, y) = E(ix, y) + iE(x, y).$$

Let $H$ corresponds to a given $E$. Then one can verify directly that the functions

$$f_\gamma(z) = \frac{1}{2i} H(z, \gamma) + \beta_\gamma$$

are the only linear solutions of (3.4), holomorphic in $z$, modulo coboundaries. Substituting in (3.3), we get the condition:

$$\frac{1}{2} H(\gamma, \gamma') + i\beta_\gamma + i\beta_{\gamma'} - i\beta_{\gamma+\gamma'} \in i\mathbb{Z}$$

for all $\gamma, \gamma' \in \Gamma$. Writing $i\beta_\gamma = \theta_\gamma + \frac{1}{4} H(\gamma, \gamma)$ we have:

$$\theta_\gamma + \theta_{\gamma'} - \theta_{\gamma+\gamma'} + \frac{i}{2} E(\gamma, \gamma') \in i\mathbb{Z} .$$

We can modify $if_\gamma$ by the coboundary of a $\mathbb{C}$-linear map $u$ on $\mathbb{C}^n$, i.e. we may replace $\theta_\gamma$ by $\theta_\gamma - u(\gamma)$ with $u : \mathbb{C}^n \to \mathbb{C}$ being $\mathbb{C}$-linear. The above condition shows that $\mathrm{Re}(\theta_\gamma)$ is additive in $\Gamma$, and hence extends to an $\mathbb{R}$-linear map $\lambda : \mathbb{C}^n \to \mathbb{R}$. Modifying $\theta$ by a map $u$ with $\mathrm{Re}(u) = \lambda$, we may assume that $\theta$ is pure imaginary. Writing $\alpha(\gamma) = e^{2\pi\theta_\gamma}$ we see that $\alpha$ has to satisfy the conditions

$$|\alpha(\gamma)| = 1, \quad \alpha(\gamma + \gamma') = e^{i\pi E(\gamma,\gamma')}\alpha(\gamma)\alpha(\gamma') .$$

One can check easily that given $E$, there exists such an $\alpha$ (just define arbitrarily $\alpha(\gamma)$ with $|\alpha(\gamma)| = 1$ for $\gamma$ in a basis of the lattice $\Gamma$ and then extend $\alpha$ by the second condition). Thus we obtained the following

**Lemma 3.7** *Let $H$ be a Hermitian form on $\mathbb{C}^n$ such that if $E = Im(H)$, then $E(\Gamma \times \Gamma) \subset \mathbb{Z}$. Let $\alpha : \Gamma \to U(1) = \{z \in \mathbb{C}^* \mid |z| = 1\}$ be a map with*

$$\alpha(\gamma + \gamma') = e^{i\pi E(\gamma,\gamma')}\alpha(\gamma)\alpha(\gamma) , \quad \gamma, \gamma' \in \Gamma ,$$

*which exists for any given $H$ as above. If we put*

$$e_\gamma(z) = \alpha(\gamma)e^{\pi H(z,\gamma) + \frac{\pi}{2} H(\gamma,\gamma')} ,$$

*then $\gamma \mapsto e_\gamma$ is a 1-cocycle on $\Gamma$ with coefficients in $H^* = H^0(\mathbb{C}^n, \mathcal{O}^*)$, such that the Chern class of the associated line bundle is $E \in H^2(X, \mathbb{Z})$.*

**Definition 3.8** A pair $(\alpha, H)$ with the above properties is called *Appell-Humbert data*.

Obviously, all the Appell-Humbert data $(\alpha, H)$ form a group with the law

$$(\alpha, H).(\alpha', H') = (\alpha\alpha', H + H') .$$

For a given pair $(\alpha, H)$ we have the 1-cocycle $(e_\gamma)$. Denoting by $L(\alpha, H)$ the line bundle on $X = \mathbb{C}^n/\Gamma$ defined by the cocycle $(e_\gamma)$ we get a map

$$\mu : \{\text{Group of data } (\alpha, H)\} \to \mathrm{Pic}(X), \quad \mu(\alpha, H) = L(\alpha, H) ,$$

which is a homomorphism. We have the main result (see [Mm4], p. 20)

**Theorem 3.9 (Appell-Humbert)** *Any line bundle $L$ on the complex torus $X$ is isomorphic to an $L(\alpha, H)$ for a uniquely determined pair $(\alpha, H)$ satisfying the conditions above. Moreover, we have the following isomorphic exact sequences:*

$$
\begin{array}{ccccccccc}
0 \to \mathrm{Hom}(\Gamma, U(1)) & \to & \left\{ \begin{array}{c} \text{Group of data} \\ (\alpha, H) \end{array} \right\} & \to & \left\{ \begin{array}{c} \text{Group of Hermitian } H \\ \text{with Im } H(\Gamma \times \Gamma) \subset \mathbb{Z} \end{array} \right\} & \to 0 \\
\lambda \downarrow \wr & & \mu \downarrow \wr & & \nu \downarrow \wr & \\
0 \to \mathrm{Pic}_0(X) & \longrightarrow & \mathrm{Pic}(X) & \longrightarrow & \mathrm{NS}(X) & \longrightarrow 0
\end{array}
$$

*where $Pic(X)$ is the group of line bundles on $X$, $Pic_0(X)$ is the subgroup of those line bundles which are topologically trivial and the last vertical map is given by $H \mapsto Im\ H$ (with the usual identification of $H^2(X, \mathbb{Z})$ with $Alt^2_{\mathbb{Z}}(\Gamma, \mathbb{Z})$).*

By the above results we have that the first row is exact and $\nu$ is an isomorphism. To finish the proof it suffices to show that $\lambda$ is an isomorphism. If $\alpha \in Hom(\Gamma, U(1))$ with $\lambda(\alpha) = 1$, then there exists $g \in H^* = H^0(\mathbb{C}^n, \mathcal{O}^*)$ such that

$$g(z + \gamma) = \alpha(\gamma)g(z) ,$$

i.e. $(e_\gamma)$ is a coboundary. If $K$ is a compact set in $\mathbb{C}^n$ with $K + \Gamma = \mathbb{C}^n$, it follows that for any $z \in \mathbb{C}^n$

$$|g(z)| \leq \sup_K |g(z)| ,$$

since $|\alpha(\gamma)| = 1$. Hence $g$ is a constant, which shows that $\lambda$ is injective ($\alpha = 1$). Consider the sequence

$$H^1(X, \mathbb{C}) \to H^1(X, \mathcal{O}_X) \to Pic_0(X) .$$

Since the first map is surjective by Hodge decomposition and, since the second map is surjective by exponential sequence, it follows that every line bundle $L \in Pic_0(X)$ is given as a quotient of the trivial line bundle $\mathbb{C}^n \times \mathbb{C}$ by an action of $\Gamma$ of the form $\Phi_\gamma(z, t) = (z + \gamma, \alpha(\gamma)t)$, where $\alpha : \Gamma \to \mathbb{C}^*$ is a homomorphism (cocycle of constant functions). As we saw above, we can normalize such actions by an automorphism of $\mathbb{C}^n \times \mathbb{C}$, such that $Im(\alpha) \subset U(1)$. Therefore $\lambda$ is surjective.

*Remark.* This fundamental result on holomorphic line bundles over complex tori is used to compute the cohomology groups of line bundles over tori and, in particular, to describe the sections of a line bundle (i.e. the so-called *theta functions*).

Let us give only the statement of the following main result (see, for instance [Mm4], p. 29):

**Theorem 3.10 (Lefschetz)** *Let $X$ be a complex torus $\mathbb{C}^n / \Gamma$, $H$ a hermitian form on $\mathbb{C}^n$ such that $E = Im(H)$ is integral on $\Gamma \times \Gamma$, $\alpha$ a map $\Gamma \to U(1)$ with $\alpha(\gamma + \gamma') = \alpha(\gamma)\alpha(\gamma')e^{i\pi E(\gamma, \gamma')}$ and $L = L(\alpha, H)$ the associated line bundle over $X$. Then the following statements are equivalent:*

*(1) The hermitian form $H$ is positive-definite.*

*(2) The space of holomorphic sections (theta functions) of $L^{\otimes n}$ gives an imbedding of $X$ as a closed complex submanifold in a projective space, for each $n \geq 3$.*

## 3.3 Neron-Severi group for some elliptic surfaces

In this section we shall study the Neron-Severi group for some elliptic surfaces, namely for elliptic bundles over a curve and for non-kählerian elliptic surfaces (see [Br2, Br5]).

We recall (see Section 2.3) that by an elliptic bundle $X \to B$, we mean that $X$ is a principal fibre bundle over a (complex, compact, connected, smooth) curve $B$, whose typical fibre and structure group are an elliptic curve $E$. The translation group $E$ is described by the universal covering sequence:

$$0 \to \Gamma \overset{j}{\to} \mathbb{C} \to E \to 0 \,, \quad \Gamma \cong \mathbb{Z}^2 \,. \tag{3.5}$$

If $B$ is any smooth, compact, connected curve, then the elliptic bundles over $B$ with typical fibre $E$ are classified by the cohomology group $H^1(B, \mathcal{E}_B)$, where $\mathcal{E}_B$ is the sheaf of germs of local holomorphic maps from $B$ to $E$, and we have the exact cohomology sequence:

$$H^1(B, \Gamma) \to H^1(B, \mathcal{O}_B) \to H^1(B, \mathcal{E}_B) \overset{c}{\to} H^2(B, \Gamma) \to 0 \,, \tag{3.6}$$

induced by (3.5).

Let $f : X \to B$ be an elliptic bundle over the curve $B$ whose typical fibre is the elliptic curve $E$. Then, the elliptic surface $X$ can be expressed in the form: $X = (B \times E)^\xi$ for some $\xi \in H^1(B, \mathcal{E}_B)$ (here we followed the notation of Kodaira [Kd1]). We need the following result:

**Proposition 3.11** *Let $f : X \to B$ be an elliptic bundle over a curve $B$. Then we have the isomorphisms: $f_* \mathbb{Z}_X \cong \mathbb{Z}_B, R^1 f_* \mathbb{Z}_X \cong \mathbb{Z}_B^2, R^2 f_* \mathbb{Z}_X \cong \mathbb{Z}_B$, and $f_* \mathcal{O}_X \cong \mathcal{O}_B, R^1 f_* \mathcal{O}_X \cong \mathcal{O}_B$.*

*Proof.* Since the fibres of $f$ are compact and connected it follows that $f_* \mathbb{Z}_X \cong \mathbb{Z}_B$. The elliptic fibration $X$ has no singular fibres, hence the local monodromy is trivial. Then the functional invariant $J$ of $X$ is constant. In fact, the surface $X$ has the same invariants as the product $B \times E$ (see Kodaira [Kd1]). It follows that the homological invariant of $X$ is trivial too (the global monodromy is trivial), hence the locally constant sheaf $R^1 f_* \mathbb{Z}_X$ is constant, i.e. $R^1 f_* \mathbb{Z}_X \cong \mathbb{Z}_B^2$. Because in each group $H^2(X_b, \mathbb{Z}) \cong \mathbb{Z}$ we have distinguished basis ($X_b = E$ is orientated) it follows that $R^2 f_* \mathbb{Z}_X \cong \mathbb{Z}_B$.

Since $f$ is connected we have $f_* \mathcal{O}_X \cong \mathcal{O}_B$. By Corollary 11.2 in [B-P-V]; Chapter III, 11, the sheaf $R^1 f_* \mathcal{O}_X$ is locally free (of rank one). Since translations operate trivialy on $H^1(E, \mathcal{O}_E)$, the line bundle $R^1 f_* \mathcal{O}_X$ on $B$ is trivial.

Now we consider the following two Leray spectral sequences (see [Go], Chapter II, 4.17):

$$E_2^{pq} = H^p(B, R^q f_* \mathbb{Z}_X) \Longrightarrow H^{p+q}(X, \mathbb{Z}), \tag{3.7}$$

$$\tilde{E}_2^{pq} = H^p(B, R^q f_* \mathcal{O}_X) \Longrightarrow H^{p+q}(X, \mathcal{O}_X). \tag{3.8}$$

Using the Proposition 3.11 and the computation of the differentials in the first spectral sequence $E_2^{pq}$ ($d_r = 0$ for $r \geq 3$) we obtain:

$$E_\infty^{02} = E_3^{02} = \mathrm{Ker}(H^0(B, R^2 f_* \mathbb{Z}_X) \overset{\alpha}{\to} H^2(B, R^1 f_* \mathbb{Z}_X)) \cong$$

$$\cong \mathrm{Ker}(H^0(B, \mathbb{Z}) \to H^2(B, \mathbb{Z}^2)), \tag{3.9}$$

where (with natural identifications) the morphism $\alpha$ is given by the multiplication with $c(\xi)$;

$$E_\infty^{11} = E_3^{11} = E_2^{11} = H^1(B, R^1 f_* \mathbb{Z}_X) \cong H^1(B, \mathbb{Z}^2); \tag{3.10}$$

$$E_\infty^{20} = E_3^{20} = \mathrm{Coker}(H^0(B, R^1 f_* \mathbb{Z}_X) \overset{\beta}{\to} H^2(B, f_* \mathbb{Z}_X)) \cong$$
$$\cong \mathrm{Coker}(H^0(B, \mathbb{Z}^2) \to H^2(B, \mathbb{Z})). \tag{3.11}$$

In the same way, for the second spectral sequence $\tilde{E}_2^{pq}$ $(d_r = 0$ for $r \geq 2)$ we have:

$$\tilde{E}_\infty^{02} = 0 \ , \ \ \tilde{E}_\infty^{20} = 0 \ ; \tag{3.12}$$

$$\tilde{E}_\infty^{11} = \tilde{E}_2^{11} = H^1(B, R^1 f_* \mathcal{O}_X) \cong H^1(B, \mathcal{O}_B). \tag{3.13}$$

Using these two spectral sequences we shall describe the kernel of the natural homomorphism $H^2(X, \mathbb{Z}) \to H^2(X, \mathcal{O}_X)$, i.e. the Neron-Severi group of the elliptic surface $X$.

**Proposition 3.12** *Let $f : X \to B$ be an elliptic bundle over a curve $B$ with typical fibre $E$, defined by $\xi \in H^1(B, \mathcal{E}_B)$ with $c(\xi) \neq 0$. Then the group $NS(X)/Tors\,NS(X)$ is isomorphic with the kernel of the natural homomorphism*

$$H^1(B, R^1 f_* \mathbb{Z}_X) \to H^1(B, R^1 f_* \mathcal{O}_X).$$

*Proof.* Let

$$0 \subset \tilde{F}_2 \subset \tilde{F}_1 \subset \tilde{F}_o = H^2(X, \mathcal{O}_X)$$

be the filtration induced by the second Leray spectral sequence (3.8). By (3.12) and (3.13) we have: $\tilde{F}_2 = 0$ and $\tilde{F}_0 = \tilde{F}_1 \cong H^1(B, R^1 f_* \mathcal{O}_X)$. It follows that

$$H^2(X, \mathcal{O}_X) \cong H^1(B, R^1 f_* \mathcal{O}_X) \ \ (\cong H^1(B, \mathcal{O}_B)).$$

Let $g$ be the genus of the curve $B$. Then $b_1(X) = 2g + 1$ is odd (see [B-P-V], p. 145 or Section 2.3), hence $X$ is nonalgebraic (with algebraic dimension $a(X) = 1$). Let

$$0 \subset F_2 \subset F_1 \subset F_0 = H^2(X, \mathbb{Z})$$

be the filtration induced by the first spectral sequence. By (3.10) we obtain:

$$F_1/F_2 \cong H^1(B, R^1 f_* \mathbb{Z}_X) \ \ (\cong H^1(B, \mathbb{Z}^2)).$$

It follows that $rk(F_1/F_2) = 4g$. Since $b_2(X) = 4g$ and since $b_2(X) \geq rk(F_1/F_2)$ we obtain that $F_2$ is a torsion subgroup of $H^2(X, \mathbb{Z})$ and $F_0 = F_1 \cong H^2(X, \mathbb{Z})$ (by (3.9) $F_0/F_1$ is included in $H^0(B, \mathbb{Z}) \cong \mathbb{Z}$, so it is torsion-free, hence it is zero).

The Neron-Severi group $NS(X)$ is the kernel of the map

$$H^2(X, \mathbb{Z}) \overset{?}{\to} H^2(X, \mathcal{O}_X)$$

induced by the natural map $\mathbb{Z}_X \to \mathcal{O}_X$. By functoriality of the spectral sequences and by the considerations above we get that the quotient group $NS(X)/TorsNS(X)$ is isomorphic with the kernel of the natural morphism

$$H^1(B, R^1 f_* \mathbb{Z}_X) \to H^1(B, R^1 f_* \mathcal{O}_X) \qquad (3.14)$$

(since $F_2 \xrightarrow{\gamma} \tilde{F}_2 = 0$, we have always $F_2 \subset NS(X)$, i.e. also in the case $c(\xi) = 0$.)

*Remark.* Using the exact sequence of small terms of the first spectral sequence (3.7)

$$0 \to H^1(B, f_* \mathbb{Z}_X) \to H^1(X, \mathbb{Z}) \to H^0(B, R^1 f_* \mathbb{Z}_X) \to$$
$$\to H^2(B, f_* \mathbb{Z}_X) \to H^2(X, \mathbb{Z}) \qquad (3.15)$$

and the formula (3.11) we have

$$F_2 \cong \mathrm{Im}(H^2(B, \mathbb{Z}) \to H^2(X, \mathbb{Z})). \qquad (3.16)$$

But $H^2(B, \mathbb{Z}) \cong NS(B)$, so the subgroup $F_2$ of the Neron-Severi group of $X$ is the image in $NS(X)$ of the Neron-Severi group of $B$ (compare [B-P-V]; Chapter IV, Theorem 2.13 or [B-F3]; Theorem 2).

We have the following result:

**Theorem 3.13** *Let $f : X \to B$ be an elliptic bundle over a curve $B$ with typical fibre $E$, defined by $\xi \in H^1(B, \mathcal{E}_B)$ with $c(\xi) \neq 0$. Then the group $NS(X)/Tors\, NS(X)$ is isomorphic with the group $\mathrm{Hom}(J_B, E)$ of the homomorphisms of abelian varieties, where $J_B$ is the Jacobian variety of the curve $B$.*

*Proof.* Using the isomorphism between the Jacobian variety and the Picard variety for a curve (see [Fo]; Theorem 21.7) and using natural identifications we obtain from the Proposition 3.12 the following isomorphism:

$$NS(X)/\mathrm{Tors}\, NS(X) \cong \mathrm{Ker}(H^1(B, \Gamma) \to H^1(B, \mathcal{O}_B)) . \qquad (3.17)$$

The exact sequence (3.5) defines the exact sequence of sheaves

$$0 \to \Gamma \to \mathcal{O}_B \to \mathcal{E}_B \to 0, \qquad (3.18)$$

and this sequence induces the exact cohomology sequence:

$$0 \to H^0(B, \Gamma) \to H^0(B, \mathcal{O}_B) \to H^0(B, \mathcal{E}_B) \to$$
$$\to H^1(B, \Gamma) \to H^1(B, \mathcal{O}_B) . \qquad (3.19)$$

From this sequence we get the isomorphisms:

$$NS(X)/\mathrm{Tors}\, NS(X) \cong \mathrm{Im}(H^0(B, \mathcal{E}_B) \to H^1(B, \Gamma)) \cong$$
$$\cong \mathrm{Coker}(H^0(B, \mathcal{O}_B) \to H^0(B, \mathcal{E}_B)). \qquad (3.20)$$

But $H^0(B, \mathcal{E}_B)$ is the group of global holomorphic maps $B \to E$ and

$$\mathrm{Im}(H^0(B, \mathcal{O}_B) \to H^0(B, \mathcal{E}_B))$$

is the subgroup of constant maps $B \to E$, which can be identified with the points of $E$. Let $B \to J_B$ be the canonical holomorphic map (determined up to a translation of $J_B$ ). Given any holomorphic map $B \to E$ ($E$ is a complex torus) then, if we choose the proper origin on $E$ (a point of $E$ "=" a constant map $B \to E$ ), the holomorphic map $B \to E$ is the composition of the canonical map $B \to J_B$ and a homomorphism from $J_B$ into $E$ (the universality of the Jacobian). It follows the desired isomorphism:

$$\mathrm{NS}(X)/\mathrm{Tors}\ \mathrm{NS}(X) \cong \mathrm{Hom}(J_B, E) .$$

Now suppose that $j(\Gamma) = \mathbb{Z} \oplus \tau\mathbb{Z}$, $\mathrm{Im}(\tau) > 0$ , and let

$$(\Pi_1, \Pi_2, ..., \Pi_{2g})$$

be a period matrix of the curve $B$. Denote by $\Lambda$ the lattice in $\mathbb{C}^g$ defined by this period matrix. Then $J_B$ is isomorphic with the quotient $\mathbb{C}^g/\Lambda$ .

**Corollary 3.14** *Let $f : X \to B$ be an elliptic bundle over a curve $B$ with typical fibre $E$ , defined by $\xi \in H^1(B, \mathcal{E}_B)$ with $c(\xi) \neq 0$ . Then we have the isomorphism*

$$\mathit{NS}(X)/\mathit{Tors}\ \mathit{NS}(X) \cong \{h : \mathbb{C}^g \to \mathbb{C} \mid h\ \mathbb{C} - \mathit{linear},\ h(\Lambda) \subset j(\Gamma)\}.$$

*Proof.* Let $X_1 = \mathbb{C}^n/\Gamma_1$, $X_2 = \mathbb{C}^m/\Gamma_2$ be two complex tori. Every analytic homomorphism $g : X_1 \to X_2$ lifts to a complex-analytic homomorphism $h : \mathbb{C}^n \to \mathbb{C}^m$, but such $h$'s are complex linear maps. Conversely, a complex linear map $h : \mathbb{C}^n \to \mathbb{C}^m$ induces an analytic homomorphism $g : X_1 \to X_2$ if and only if $h(\Gamma_1) \subset \Gamma_2$ (see [Mm4], p. 175). In particular, we get the result.

**Corollary 3.15** *Let $f : X \to B$ be an elliptic bundle over a curve $B$ with typical fibre $E$ , defined by $\xi$ with $c(\xi) \neq 0$ . Let $r$ be the natural number*

$$rk_{\mathbb{Q}}\{\Pi_1, ..., \Pi_{2g}, \tau\Pi_1, ..., \tau\Pi_{2g}\}.$$

*Then the rank of the Neron-Severi group $NS(X)$ is given by $\rho(X) = 4g - r$ .*

*Proof.* The image of the map $H^1(B, \Gamma) \to H^1(B, \mathcal{O}_B)$ is generated by the vectors $\Pi_1, ..., \Pi_{2g}, \tau\Pi_1, ..., \tau\Pi_{2g}$, identifying as usual the Jacobian variety with the Picard variety of $B$. It follows that $\rho(X) = \mathrm{rk}(\mathrm{NS}(X)) = 4g - r$.

*Remark.* Obviously, $2g \leq r \leq 4g$ ; hence $0 \leq \rho(X) \leq 2g$ .

**Examples (1)** If $B = \mathbb{P}^1$ then $X$ is a Hopf surface (see [B-P-V]; Chapter V, Theorem 5.4) and $\rho(X) = b_2(X) = 0$ .
**(2)** Now, let $B$ be also an elliptic curve and suppose that the lattice $\Lambda = \mathbb{Z} \oplus \varepsilon\mathbb{Z}$, $\mathrm{Im}(\varepsilon) > 0$. Then $X$ is a primary Kodaira surface (see Section 2.3) and $b_2(X) = 4$ , but $0 \leq \rho(X) \leq 2$ . Choosing conveniently $\tau$ and $\varepsilon$ one sees, easily, that $\rho(X)$ takes all the values 0, 1 or 2. Of course, in this case, $r = \mathrm{rk}_{\mathbb{Q}}\{1, \tau, \varepsilon, \tau\varepsilon\}$ .
**(3)** Let us suppose that $g \geq 2$ . From Proposition 12.5 [B-P-V]; Chapter V we have

that $X$ is a minimal properly elliptic surface (nonalgebraic and, even, non-Kähler, since $b_1(X)$ is odd).

*Remark.* In the case $c(\xi) = 0$, the surface $X$ can be nonalgebraic (see Kodaira [Kd2]), but it is Kähler (see [Mi]); the subgroup $F_2$ of the Neron-Severi group $NS(X)$ is no longer a torsion subgroup. In fact, $F_2$ is generated by $c_1(E) \neq 0$, which is not a torsion element in $NS(X)$ because $X$ is Kähler.

We have:

**Theorem 3.16** *Let $f : X \to B$ be an elliptic bundle over a curve $B$ with typical fibre $E$, defined by $\xi \in H^1(B, \mathcal{E}_B)$ with $c(\xi) = 0$. Let $r$ be $rk_\mathbb{Q}\{\Pi_1, ..., \Pi_{2g}, \tau\Pi_1, ..., \tau\Pi_{2g}\}$. Then the rank of the Neron-Severi group $NS(X)$ is given by $\rho(X) = 4g - r + 1 + \theta$, where $\theta = 0$ or $1$.*

*Proof.* The contribution of the subgroup $F_2$ to the rank of the Neron-Severi group $NS(X)$ is one in this case. From the proof of Proposition 3.12 it follows that the contribution of $F_1/F_2$ to the rank of the Neron-Severi group $NS(X)$ is given by the rank of the kernel of the natural morphism (3.14). As in the proofs of Theorem 3.13 and Corollary 3.15 we get that this rank is equal with $4g - r$. Finally, we obtain that the rank of the Neron-Severi group $NS(X)$, in this case, is $4g - r + 1 + \theta$, where $\theta = 0$ or $1$, and represents the contribution of $F_0/F_1$.

*Remark.* In the case $c(\xi) = 0$, we have: $1 \leq \rho(X) \leq 2g + 1 + \theta$.

**Examples (1)** If $B = \mathbb{P}^1$ and $X = B \times E$, then $\rho(X) = 2$ (and $\theta = 1$).
**(2)** Let $X$ be a two dimensional complex torus of algebraic dimension one. One knows (see [E-F]; Appendix) that, in this case, $1 \leq \rho(X) \leq 3$. An algebraic reduction gives it the structure of an elliptic bundle over an elliptic curve $B$, which can be defined by locally constant transition functions ( $c(\xi) = 0$ cf. [B-P-V]; Chapter V.5). Of course, in this case, we have $\theta = 0$, if $\rho(X) = 1$.

Now, we shall extend the results for non-kählerian elliptic surfaces (see [Br5]). Let $\varphi : X \to B$ be an elliptic surface. We shall always assume that $\varphi$ is relatively minimal, i.e. all fibres are free of $(-1)$-curves.
Let $F = \sum n_i D_i$ be a singular fibre of $\varphi$, where $D_i$ s are the irreducible reduced components and the $n_i$ s are their multiplicities. Let $m$ denotes the greatest common divisor of the $n_i$ s. If $m \geq 2$, then the fibre $F$ is called *multiple fibre of multiplicity* $m$ and we will write $F = mD$, where $D = \sum (n_i/m)D_i$.
An elliptic surface $\varphi : X \to B$ is called a *quasi-bundle* if all smooth fibres are pairwise isomorphic, and the only singular fibres are multiples of smooth (elliptic) curves. If moreover $\varphi$ has no singular fibres then $\varphi : X \to B$ is said to be a *fibre bundle*.

Let $\varphi : X \to B$ be a non-kählerian elliptic surface. We need the following (generally) known result (for algebraic elliptic surfaces see, for example, [So2]):

**Proposition 3.17** *A non-kählerian elliptic surface $\varphi : X \to B$ is a quasi-bundle.*

*Proof.* In view of a result of Miyaoka (see [Mi]) the first Betti number is odd. By the Theorem 2.6 of Chapter IV in [B-P-V] one has $b_1(X) = 2q(X) - 1$, where $q(X) = h^{0,1} = \dim H^1(X, \mathcal{O}_X)$ is the irregularity of $X$. Denote by $m_1 D_1, ..., m_t D_t$ all multiple fibres of $\varphi$ and let $E$ be any smooth fibre. Define

$$G(\varphi) := \operatorname{Coker}(\mathbb{Z} \to \oplus \mathbb{Z}_{m_i}), \quad 1 \to (1, ..., 1).$$

Then there exists an exact sequence

$$H_1(E, \mathbb{Z}) \to H_1(X, \mathbb{Z}) \to H_1(B, \mathbb{Z}) \times G(\varphi) \to 0 , \tag{3.21}$$

induced by $\varphi$ and the inclusion of $E$ in $X$ (see [Sol], Theorem 1.3). Since $b_1(X)$ is odd and rank $H_1(E, \mathbb{Z}) = 2$, it follows that the rank of the image of $H_1(E, \mathbb{Z})$ is one. We get $b_1(X) = 2g(B) + 1$, hence $q(X) - g(B) = 1$ ($g(B)$ is the genus of the curve $B$). The first terms of the Leray spectral sequence

$$E_2^{p,q} = H^p(B, R^q \varphi_* \mathcal{O}_X) \Longrightarrow H^{p+q}(X, \mathcal{O}_X)$$

yield the exact sequence

$$0 \to H^1(B, \varphi_* \mathcal{O}_X) \to H^1(X, \mathcal{O}_X) \to H^0(B, R^1 \varphi_* \mathcal{O}_X) \to 0 \tag{3.22}$$

Since $\varphi_* \mathcal{O}_X \cong \mathcal{O}_B$ we get

$$h^0(B, R^1 \varphi_* \mathcal{O}_X) = q(X) - g(B) = 1 \tag{3.23}$$

and

$$\chi(\mathcal{O}_X) = \deg(R^1 \varphi_* \mathcal{O}_X)^*$$

(see [B-P-V], Chapter V, Proposition 12.2).By relative duality (see [B-P-V], p. 99), one has:

$$\varphi_*(\omega_{X/B}) \cong (R^1 \varphi_* \mathcal{O}_X)^*,$$

where " $*$ " denotes dual as $\mathcal{O}_B$-module. The Theorem 18.2, Chapter III, in [B-P-V] shows that

$$\deg(R^1 \varphi_* \mathcal{O}_X)^* = \deg(\varphi_*(\omega_{X/B})) \geq 0$$

and this degree vanishes if and only if all the smooth fibres of $\varphi$ are isomorphic and the singular fibres are of type $m I_0$, i.e. multiples of smooth (elliptic) curves. Suppose $\deg(\varphi_*(\omega_{X/B})) > 0$. Since $\deg(R^1 \varphi_* \mathcal{O}_X) < 0$, then $h^0(B, R^1 \varphi_* \mathcal{O}_X) = 0$ and we get a contradiction with (3.23). It follows that $\chi(\mathcal{O}_X) = 0$ and $\varphi : X \to B$ is a quasi-bundle.

**Lemma 3.18** *Let* $\varphi : X \to B$ *be a quasi-bundle with the first Betti number odd. Let* $m_1 D_1, ..., m_t D_t$ *be all multiple fibres of* $\varphi$ *and let* $m$ *denote the least common multiple of* $m_1, ..., m_t$. *Then there exist an elliptic bundle* $\psi : Y \to C$, *with the first Betti number odd, and two cyclic coverings* $\varepsilon : C \to B, \pi : Y \to X$, *both with group* $\mathbb{Z}_m$, *such that* $\varphi \circ \pi = \varepsilon \circ \psi$.

*Proof.* Choose an integer $e \geq 0$ such that $m$ divides $t + e$. Let $\varphi(D_i) = P_i \in B$, $i = 1, ..., t$, and take distinct points $P_{t+1}, ..., P_{t+e} \in B$, which are different from $P_i, i = 1, ..., t$. Then there is at least one line bundle $L$ on $B$ with

$$L^{\otimes m} \cong \mathcal{O}_B(P_1 + ... + P_{t+e}).$$

Such an $L$ defines a cyclic covering $\varepsilon : C \to B$ of degree $m$, totally ramified at $P_1, ..., P_{t+e}$ (see [B-P-V],Chapter I, Lemma 17.1 or Section 2.3). Let $Y$ be the normalization of $X \times_B C$. Then $Y$ is smooth by [B-P-V], Chapter III, Proposition 9.1, and there exists a cyclic covering map $\pi : Y \to X$ with group $G \cong \mathbb{Z}_m$, i.e. $X = Y/G$. We get the following commutative diagram:

$$
\begin{array}{ccc}
Y & \xrightarrow{\pi} & X \\
\psi \downarrow & & \downarrow \varphi \\
C & \xrightarrow{\varepsilon} & B
\end{array}
$$

For this construction see [So2] or [B-P-V], Chapter III, Theorem 10.3. By [B-P-V], Chapter III, Proposition 9.1, one sees that $\psi : Y \to C$ is a fibre bundle. Denote by $E$ a smooth fibre of $\varphi$ and by $\tilde{E}$ a connected component of $\pi^{-1}(E)$. Then $\tilde{E}$ is a fibre of $\psi$ and the restriction $\tilde{E} \to E$ of $\pi$ is an isomorphism. Now, we can apply to this situation the results of Kodaira in [Kd1, Kd2]. The elliptic surface $Y$ has no singular fibres, hence the local monodromy is trivial. Then the functional invariant $J$ of $Y$ is constant. Since $X = Y/G$ is not a deformation of an algebraic surface, by Theorems 14.5 and 14.6 in [Kd1], we deduce that the homological invariant of $Y$ is trivial too (the global monodromy is trivial). It follows that $\psi : Y \to C$ is an elliptic bundle defined by an element $\xi \in H^1(C, \mathcal{E}_C)$. From the proof of the Theorem 14.7 in [Kd1] we get $c(\xi) \neq 0$ hence, by the Theorem 11.9 in [Kd1] (or, by the Proposition 5.3, Chapter V, in [B-P-V]), we obtain that $b_1(Y)$ is odd.

Let $\varphi : X \to B$ be a non-kählerian elliptic surface. From the previous results we get the diagram:

$$
\begin{array}{ccc}
Pic(X) & \xrightarrow{c_1} & H^2(X, \mathbb{Z}) \\
\pi^* \downarrow & & \downarrow \pi^* \\
Pic(Y) & \xrightarrow{c_1} & H^2(Y, \mathbb{Z})
\end{array}
$$

Let $G \cong \mathbb{Z}_m$ be the finite group of the cyclic covering $\pi : Y \to X$. We need the following result:

**Lemma 3.19** *If* $\pi : Y \to X$ *is a flat cyclic covering with* $X$ *non-singular and* $Y$ *normal then*

$$Pic(X) \otimes \mathbb{Q} \cong Pic(Y)^G \otimes \mathbb{Q} \quad and \quad NS(X) \otimes \mathbb{Q} \cong NS(Y)^G \otimes \mathbb{Q},$$

where $Pic(Y)^G$ (resp. $NX(Y)^G$) is the subgroup of invariants of the Picard group $Pic(Y)$ (resp. Neron-Severi group $NS(Y)$).

*Proof.* Since $c_1$ is compatible with the group action, the first statement implies the second.Obviously, $\pi^*Pic(X) \subset Pic(Y)^G$. If $\mathcal{N} \in Pic(Y)^G$, then $(\pi_*(\mathcal{N}))^G$ is a line bundle. Indeed, because $\mathcal{N}$ is a $G$-sheaf we have an action of $G$ on $\pi_*(\mathcal{N})$. But $G$ is an abelian group, so the sheaf of invariants $\pi_*(\mathcal{N})^G$ of $\pi_*(\mathcal{N})$ is a line bundle (the invariant summand in the splitting of $\pi_*(\mathcal{N})$ as a direct sum according to the characters of $G$); see [Mm4], p. 72, [Ct], or, for a systematic study of abelian covers [Pa]. Now, if $m$ is the order of $G$, then one has

$$\pi^*(\pi_*\mathcal{N}^m)^G \cong \mathcal{N}^m.$$

We have a natural morphism of line bundles

$$\pi^*(\pi_*(\mathcal{N}^m)^G) \to \mathcal{N}^m,$$

so, in order to obtain the isomorphism above, it is sufficient to show such an equality in a neighbourhood of $y \in Y$. One chooses a Stein neighbourhood $V$ of $x = \pi(y)$; $\pi^{-1}(V)$ is Stein and one finds $s \in \mathcal{N}(\pi^{-1}(V))$ with $s(y_i) \neq 0$ for $y_i \in \pi^{-1}(x)$. Then $\prod_{g \in G} s^g$ is a $G$-invariant section of $\mathcal{N}^m(\pi^{-1}(V))$ which generates $\mathcal{N}^m$ in a neighbourhood of $\pi^{-1}(x)$.

**Lemma 3.20** *Let $\varepsilon : C \to B$ be a cyclic covering (with group $G \cong \mathbb{Z}_m$) of curves and let $E$ be an elliptic curve. Then there exists an exact sequence of groups*

$$0 \to Hom(J_B, E) \to Hom(J_C, E)^G \to Hom(G, E),$$

*where $J_B$, resp. $J_C$, is the Jacobian variety of the curve $B$, resp. $C$.*

*Proof.* (N. Buruiană) By a suitable choice we can suppose that we have the diagram:

$$
\begin{array}{ccc}
C & \hookrightarrow & J_C \\
\varepsilon \downarrow & & \downarrow \varepsilon \\
B & \hookrightarrow & J_B
\end{array}
$$

Let $h$ be the genus of the curve $C$ and let $\alpha = x_1 + ... + x_h \in J_C$, where $x_1, ..., x_h \in C$. If $f \in Hom(J_C, E)^G$, then for all $g \in G$

$$f(gx_1 + ... + gx_h) = f(x_1 + ... + x_h).$$

Take $x_1 = x \in C$ an arbitrary element and $x_2 = ... = x_h = 0 \in C \subset J_C$. Then we get

$$f(gx + (h-1)g0) = f(x) = \tilde{f}(x),$$

where $\tilde{f}$ is the restriction of $f$ to $C$. It follows

$$\tilde{f}(gx) - \tilde{f}(x) = a_g \in E,$$

for all $x \in C$. Thus we obtain a map $(g \to a_g) : G \to E$. But

$$\tilde{f}(gg'x) - \tilde{f}(x) = \tilde{f}(gg'x) - \tilde{f}(g'x) + \tilde{f}(g'x) - \tilde{f}(x),$$

i.e.

$$a_{gg'} = a_g + a_{g'},$$

and so $(g \to a_g)$ is a homomorphism $a^f : G \to E$. Clearly, $a^f = 0$ if and only if $\tilde{f}(gx) = \tilde{f}(x)$ for all $g \in G$ and all $x \in C$, i.e. if and only if there exists $\tilde{u} : B \to E$ such that $\tilde{f} = \tilde{u} \circ \varepsilon$. But the elements of a curve generates (as a group) the corresponding Jacobian, so we have $f = u \circ \varepsilon$, where $u : J_B \to E$ is a homomorphism.

**Theorem 3.21** *Let $X \to B$ be a non-kählerian elliptic surface. Then we have the isomorphism*

$$NS(X) \otimes \mathbb{Q} \cong Hom(J_B, E) \otimes \mathbb{Q},$$

*where $J_B$ is the Jacobian variety of the curve $B$ and $Hom(J_B, E)$ is the group of the homomorphisms of abelian varieties.*

*Proof.* By the previous results and the Theorem 3.13.

*Remark.* These results were extended and precised for the case of torus quasi-bundles over curves (see [B-U]).

## 3.4 Picard group for primary Kodaira surfaces

In this section we shall give an explicit description of the Picard group $Pic(X)$ of a primary Kodaira surface $X$, similar to the Appell-Humbert Theorem for complex tori (see [Br3]).

Recall (see Section 2.3) that a primary Kodaira surface is an elliptic bundle $X \to B$ over an elliptic curve $B$ defined by a class $\xi \in H^1(B, \mathcal{E}_B)$ with $c(\xi) \neq 0$. Its invariants are

$$H^1(X, \mathbb{Z}) \cong \mathbb{Z}^3 ; H^2(X, \mathbb{Z}) \cong \mathbb{Z}^4 \oplus \mathbb{Z}_m, m \in \mathbb{N}^* ;$$

$$e(X) = 0; \ h^1(\mathcal{O}_X) = 2; \ h^2(\mathcal{O}_X) = 1; \ \mathcal{K}_X \cong \mathcal{O}_X;$$

(see, for example, [B-P-V], Chapter V.5).

We need a more concrete description of a primary Kodaira surface (see, for example, [Kd2]). The universal covering space of X is $\mathbb{C}^2$ and its fundamental group is isomorphic with a non-abelian group G of affine automorphisms of $\mathbb{C}^2$ with four generators $g_1, g_2, g_3, g_4$ of the following form

$$g_i(z, w) = (z, w + \beta_i), \ i = 1, 2; \tag{3.24}$$

$$g_i(z, w) = (z + \alpha_i, w + \overline{\alpha}_i z + \beta_i), \ i = 3, 4,$$

and one relation

$$g_3 g_4 = g_1^m g_4 g_3 , \quad m \in \mathbb{N}^* . \tag{3.25}$$

The fundamental group of the fibre $E$ is isomorphic with the lattice $\Gamma \subset \mathbb{C}$ generated by $\beta_1, \beta_2$, the fundamental group of the base $B$ is isomorphic with the lattice $\Lambda \subset \mathbb{C}$ generated by $\alpha_3, \alpha_4$ and we have the following central extension

$$0 \to \Gamma \overset{j}{\to} G \overset{\pi}{\to} \Lambda \to 0 \tag{3.26}$$

From (3.25) we get the formula

$$\overline{\alpha}_3 \alpha_4 - \overline{\alpha}_4 \alpha_3 = m\beta_1. \tag{3.27}$$

Puting an element $g \in G$ in the form $g = g_4^{l_4} g_3^{l_3} g_2^{l_2} g_1^{l_1}$, with $l_i \in \mathbb{Z}, i = 1, 2, 3, 4$, we have on G the law

$$gg' = g_4^{l_4+l_4'} g_3^{l_3+l_3'} g_2^{l_2+l_2'} g_1^{l_1+l_1'+ml_3 l_4'} , \quad g, g' \in G. \tag{3.28}$$

Obviously, $j(\gamma) = g_2^{l_2} g_1^{l_1}$ for $\gamma = \beta_1 l_1 + \beta_2 l_2 \in \Gamma$ and $\pi(g) = \alpha_3 l_3 + \alpha_4 l_4 = \lambda \in \Lambda$. Choosing as a cross-section of $\pi$ the map $s : \Lambda \to G$ , $s(\lambda) = g_4^{l_4} g_3^{l_3}$ for $\lambda = \alpha_3 l_3 + \alpha_4 l_4 \in \Lambda$ , we have

$$s(\lambda)s(\lambda') = s(\lambda + \lambda')g_1^{ml_3 l_4'}, \quad \lambda, \lambda' \in \Lambda . \tag{3.29}$$

By identifying $\gamma$ with $j(\gamma)$ and taking into account that $\gamma$ commutes with any element of G we can write uniquely $g = \gamma s(\lambda)$ for any $g \in G$ . The action of an element $g \in G$ on $\mathbb{C}^2$ is given by the formula:

$$g(z, w) = (z + \lambda, w + \overline{\lambda} z + \beta(g)) , \tag{3.30}$$

where

$$\begin{cases} \beta(g) &= \gamma + \beta_1(s(\lambda)) + \beta_2(s(\lambda)) \\ \beta_1(s(\lambda)) &= \tilde{\beta}_3 l_3 + \tilde{\beta}_4 l_4 \\ \tilde{\beta}_3 &= \beta_3 - (1/2)\alpha_3 \overline{\alpha}_3 \\ \tilde{\beta}_4 &= \beta_4 - (1/2)\alpha_4 \overline{\alpha}_4 \\ \beta_2(s(\lambda)) &= (1/2)\alpha_3 \overline{\alpha}_3 l_3^2 + (1/2)\alpha_4 \overline{\alpha}_4 l_4^2 + \alpha_3 \overline{\alpha}_4 l_3 l_4. \end{cases} \tag{3.31}$$

Once we know that the universal covering space of the surface $X$ is $\mathbb{C}^2$ then there is a general method for constructing holomorphic line bundles on the surface $X$ (see, for example [Mm4], p. 13). Every line bundle $L$ on $X$ is obtained as the quotient of the trivial line bundle $\mathbb{C}^2 \times \mathbb{C}$ on $\mathbb{C}^2$ by a linear action of $G$ on the trivial line bundle covering the action of $G$ on the base $\mathbb{C}^2$ . Let us denote by $H^*$ the multiplicative group $H^0(\mathbb{C}^2, \mathcal{O}_{\mathbb{C}^2}^*)$ of non-vanishing holomorphic functions on $\mathbb{C}^2$. Then the action of $G$ on $\mathbb{C}^2 \times \mathbb{C}$ is given by:

$$\Phi_g((z, w), t) = (g(z, w), e_g(z, w)t), \ g \in G , \tag{3.32}$$

where $e_g \in H^*$ . The condition that $\Phi_g$ is a group action gives that $g \mapsto e_g$ is a 1-cocycle for the group $G$ with coefficients in $H^*$ :

$$e_{gg'}(z, w) = e_g(g'(z, w))e_{g'}(z, w) , \ g, g' \in G . \tag{3.33}$$

In fact, we get an isomorphism (see [Mm4], Chapter I, 2, Appendix):

$$\Phi \; : \; H^1(G, H^*) \cong H^1(X, \mathcal{O}_X^*) \, .$$

Now, our problem is to find the simplest kind of representing cocycles $(e_g)$ for all cohomology classes in $H^1(X, \mathcal{O}_X^*)$.

We need some well-known facts on cohomology of groups (see also [ML], Chapter XI). In fact, we shall need later the very explicit description given in the proofs. Let

$$0 \to \Gamma \xrightarrow{j} G \xrightarrow{\pi} \Lambda \to 0 \tag{3.34}$$

be a *central group extension* (i.e. $\Gamma$ is abelian and $j(\Gamma)$ is in the center of $G$) . Suppose for simplicity that $\Lambda$ is abelian too . We shall identify $\Gamma$ with its image $j(\Gamma)$ in $G$ and let $s : \Lambda \to G$ be a cross-section, i.e. $\pi \circ s = 1_\Lambda$ (where the group $\Lambda$ is identified with $G/\Gamma$ ). Then the elements of $G$ can be uniquely written in the form $\gamma s(\lambda)$, where $\gamma \in \Gamma$ and $\lambda \in \Lambda$ . The sum $s(\lambda)s(\lambda')$ must lie in the same coset as $s(\lambda + \lambda')$ , so there are unique elements $h_0(\lambda, \lambda') \in \Gamma$ such that always

$$s(\lambda)s(\lambda') = h_0(\lambda. \lambda')s(\lambda + \lambda') \, . \tag{3.35}$$

Now, consider $\mathbb{Z}$ as a trivial $G$ -module and let

$$\mathrm{res} : H^2(G, \mathbb{Z}) \to H^2(\Gamma, \mathbb{Z}) \tag{3.36}$$

be the restriction homomorphism.

**Lemma 3.22** *Let $\eta \in H^2(G, \mathbb{Z})$ be an element of the kernel of the restriction map. Then there exists a 2-cocycle $\tilde{F} : G \times G \to \mathbb{Z}$ representing $\eta$ such that*

*(a) $\tilde{F}(\gamma, \gamma') = 0$ , $\gamma, \gamma' \in \Gamma$,*

*(b) $\tilde{F}(\gamma, s(\lambda)) = 0$ , $\gamma \in \Gamma, \lambda \in \Lambda$,*

*(c) $\tilde{F}(\gamma s(\lambda), \gamma's(\lambda')) = \tilde{F}(s(\lambda), s(\lambda')) + \tilde{F}(s(\lambda), \gamma')$ , $\gamma, \gamma' \in \Gamma ; \lambda, \lambda' \in \Lambda$.*

*Moreover, the restriction of $\tilde{F}$ to $G \times \Gamma$ is bilinear and independent of the choice of $\tilde{F}$ with the properties (a) - (c) .*

*Proof.* Let $F : G \times G \to \mathbb{Z}$ be a normalized 2-cocycle representing $\eta$ and denote also by $F$ the restriction to $\Gamma \times \Gamma$ . Since $\mathrm{res}(\eta) = 0$ in $H^2(\Gamma, \mathbb{Z})$ it follows that there exists a normalized 1-cochain $\theta : \Gamma \to \mathbb{Z}$ such that

$$F(\gamma, \gamma') = \theta(\gamma) + \theta(\gamma') - \theta(\gamma + \gamma') , \; \gamma, \gamma' \in \Gamma \, . \tag{3.37}$$

Now, define the normalized 1-cochain $\Omega : G \to \mathbb{Z}$ by

$$\Omega(g) = F(\gamma, s(\lambda)) - \theta(\gamma)^*, \tag{3.38}$$

where $g = \gamma s(\lambda)$ , $\gamma \in \Gamma$ and $\lambda \in \Lambda$ . Then, the $\tilde{F} : G \times G \to \mathbb{Z}$ defined by

$$\tilde{F}(g, g') = F(g, g') + \Omega(g) + \Omega(g') - \Omega(gg') \tag{3.39}$$

for all $g, g' \in G$, is a normalized 2-cocycle cohomologous to $F$, so representing $\eta$. For any $\gamma, \gamma' \in \Gamma$, by (3.39), we have

$$\tilde{F}(\gamma, \gamma') = F(\gamma, \gamma') + F(\gamma, 0) - \theta(\gamma) + F(\gamma', 0) - \theta(\gamma') - F(\gamma + \gamma', 0) + \theta(\gamma + \gamma') = 0.$$

For any $\gamma \in \Gamma$, $\lambda \in \Lambda$ we obtain

$$\tilde{F}(\gamma, s(\lambda)) = F(\gamma, s(\lambda)) + F(\gamma, 0) - \theta(\gamma) + F(0, s(\lambda)) - \theta(0) - F(\gamma, s(\lambda)) + \theta(\gamma) = 0$$

(since $F, \theta$ are normalized). This completes the proof of (a) and (b).

Since $\tilde{F}$ is a cocycle, it follows

$$\tilde{F}(\gamma s(\lambda), \gamma' s(\lambda')) = \tilde{F}(\gamma, \gamma' s(\lambda) s(\lambda')) + \tilde{F}(s(\lambda), \gamma' s(\lambda')) - \tilde{F}(\gamma, s(\lambda)).$$

By using (3.35), (a), (b) and, again, the fact that $\tilde{F}$ is a cocycle we get that

$$\tilde{F}(\gamma s(\lambda), \gamma' s(\lambda')) = \tilde{F}(\gamma + \gamma' + h_0(\lambda, \lambda'), s(\lambda + \lambda')) - \tilde{F}(\gamma' + h_0(\lambda, \lambda'), s(\lambda + \lambda')) +$$

$$+ \tilde{F}(\gamma, \gamma' + h_0(\lambda, \lambda')) + \tilde{F}(\gamma' s(\lambda), s(\lambda')) + \tilde{F}(s(\lambda), \gamma') - \tilde{F}(\gamma', s(\lambda')).$$

Finally, we obtain

$$\tilde{F}(\gamma s(\lambda), \gamma' s(\lambda')) = \tilde{F}(s(\lambda), s(\lambda')) + \tilde{F}(s(\lambda), \gamma'),$$

hence (c) holds. By (c)

$$\tilde{F}(g, \gamma + \gamma') = \tilde{F}(g\gamma, \gamma') + \tilde{F}(g, \gamma) - \tilde{F}(\gamma, \gamma') = \tilde{F}(g, \gamma) + \tilde{F}(g, \gamma')$$

and similarly

$$\tilde{F}(gg', \gamma) = \tilde{F}(g, g'\gamma) + \tilde{F}(g', \gamma) - \tilde{F}(g, g') = \tilde{F}(g, \gamma) + \tilde{F}(g', \gamma).$$

Thus, the restriction of $\tilde{F}$ to $G \times \Gamma$ is bilinear. Assume now that there exist two cocycles $\tilde{F}_1, \tilde{F}_2 \in \eta$ with the properties (a) - (c). Then $\tilde{F}_1 - \tilde{F}_2 = \delta h$, where $h : G \to \mathbb{Z}$ is a 1-cochain. For any $g \in G$ and any $\gamma' \in \Gamma$ we have

$$(\tilde{F}_1 - \tilde{F}_2)(g, \gamma') = h(g) + h(\gamma') - h(g\gamma') = h(\gamma') + h(g) - h(\gamma'g) =$$

$$= (\tilde{F}_1 - \tilde{F}_2)(\gamma', g) = (\tilde{F}_1 - \tilde{F}_2)(0, s(\lambda)) + (\tilde{F}_1 - \tilde{F}_2)(0, \gamma) = 0$$

by (c), hence the restriction of $\tilde{F}$ to $G \times \Gamma$ is unique.

Because $\mathbb{Z}$ is a trivial $G$-module the inflation homomorphism has the form

$$\inf : H^2(\Lambda, \mathbb{Z}) \to H^2(G, \mathbb{Z}). \tag{3.40}$$

For $\Lambda$ abelian we have canonical isomorphisms:

$$H^1(\Lambda, H^1(\Gamma, \mathbb{Z})) \cong \mathrm{Hom}_{\mathbb{Z}}(\Lambda \otimes \Gamma, \mathbb{Z}) \cong \mathrm{Bil}(\Lambda \times \Gamma, \mathbb{Z}). \tag{3.41}$$

For the next result (with some different assumptions) see [ML], Chapter XI or [H-S].

**Lemma 3.23** *Suppose that the restriction map (3.36) is zero. Then there exists a homomorphism $v$ such that the sequence*

$$H^2(\Lambda, \mathbb{Z}) \overset{inf}{\to} H^2(G, \mathbb{Z}) \overset{v}{\to} H^1(\Lambda, H^1(\Gamma, \mathbb{Z})) \qquad (3.42)$$

is exact.

*Proof.* By (3.41) we can define the homomorphism $v$ in the form

$$v : H^2(G, \mathbb{Z}) \to \text{Bil}(\Lambda \times \Gamma, \mathbb{Z}) .$$

Let $\eta \in H^2(G, \mathbb{Z})$ and let $\tilde{F}$ be a 2-cocycle representing $\eta$, which satisfies the properties from the previous lemma. The restriction of $\tilde{F}$ to $G \times \Gamma$ is bilinear and independent of the choice of $\tilde{F}$. Therefore, it induces a unique bilinear map

$$\overline{F} : \Lambda \times \Gamma \to \mathbb{Z} \qquad (3.43)$$

and we shall define $v(\eta) := \overline{F}$. Obviously, $v$ is a homomorphism. Let now $\xi \in H^2(\Lambda, \mathbb{Z})$ and let $f$ be a normalized 2-cocycle representing $\xi$. Then $\inf(\xi)$ is represented by the 2-cocycle

$$F(g, g') = f(\pi(g), \pi(g')) = f(\lambda, \lambda') \ , g, g' \in G .$$

Clearly, $F$ satisfies the properties from the previous lemma. Hence, for $\lambda \in \Lambda$ and $\gamma \in \Gamma$ we get

$$v(\inf(\xi))(\lambda, \gamma) = \overline{F}(\lambda, \gamma) = f(\lambda, 0) = 0 .$$

Thus $v \circ \inf = 0$. Let now $\eta \in \text{Ker } v$ ; then $v(\eta)(\lambda, \gamma) = \overline{F}(\lambda, \gamma) = 0$ , $\lambda \in \Lambda$ , $\gamma \in \Gamma$ , hence $\tilde{F}(g, \gamma) = 0$ for any $g \in G$ and $\gamma \in \Gamma$ . It follows that

$$\tilde{F}(g, g') = \tilde{F}(s(\lambda), s(\lambda')) \ , g, g' \in G .$$

Define the 2-cocycle $f : \Lambda \times \Lambda \to \mathbb{C}$ by $f(\lambda, \lambda') := \tilde{F}(s(\lambda), s(\lambda'))$ and let $\xi \in H^2(\Lambda, \mathbb{Z})$ be its class. Obviously, $\inf(\xi) = \eta$ and the sequence (3.42) is exact.

*Remark.* This very explicit construction of the homomorphism $v$ will be used later.

Let $X \to B$ be a primary Kodaira surface. We know from Theorem 3.13, that the group $\text{NS}(X)/\text{Tors NS}(X)$ is isomorphic with the group $\text{Hom}(B, E)$ of homomorphisms of elliptic curves, where $\text{NS}(X)$ denotes the Neron-Severi group of $X$ and $\text{Tors NS}(X)$ means the torsion subgroup of $\text{NS}(X)$. In the proof of this result we used the filtration on the group $H^2(X, \mathbb{Z})$ given by the Leray spectral sequence. Now we want to compare that filtration with the filtration given by the Lyndon spectral sequence (see [ML], Chapter XI):

$$E_2^{pq} = H^p(\Lambda, H^q(\Gamma, \mathbb{Z})) \implies H^{p+q}(G, \mathbb{Z}) . \qquad (3.44)$$

This is possible because, by a well-known result (see, for example [ML] , Chapter IV, Theorem 11.5), we have natural isomorphisms for any $i$:

$$H^i(X, \mathbb{Z}) \cong H^i(G, \mathbb{Z}), \qquad (3.45)$$

$$H^i(E, \mathbb{Z}) \cong H^i(\Gamma, \mathbb{Z}) , \ H^i(B, \mathbb{Z}) \cong H^i(\Lambda, \mathbb{Z}) .$$

Similarly to the Leray spectral sequence in [Br2], since $d_r = 0$ for $r \geq 3$, we have:

$$E_\infty^{02} \cong E_3^{02} = \mathrm{Ker}(H^0(\Lambda, H^2(\Gamma, \mathbb{Z})) \to H^2(\Lambda, H^1(\Gamma, \mathbb{Z}))) , \qquad (3.46)$$

$$E_\infty^{11} \cong E_3^{11} \cong E_2^{11} = H^1(\Lambda, H^1(\Gamma, \mathbb{Z})) , \qquad (3.47)$$

$$E_\infty^{20} \cong E_3^{20} = \mathrm{Coker}(H^0(\Lambda, H^1(\Gamma, \mathbb{Z})) \to H^2(\Lambda, H^0(\Gamma, \mathbb{Z}))) . \qquad (3.48)$$

Identifying the groups $H^2(G, \mathbb{Z})$ and $H^2(X, \mathbb{Z})$, by the isomorphism (3.45), we obtain:

**Lemma 3.24** *The Lyndon spectral sequence* (3.44) *gives on* $H^2(X, \mathbb{Z})$ *the same filtration as the Leray spectral sequence does. Moreover, the restriction map is zero.*

*Proof.* Let

$$0 \subset F_2' \subset F_1' \subset F_0' = H^2(G, \mathbb{Z}) \cong H^2(X, \mathbb{Z})$$

be the filtration induced by the Lyndon spectral sequence. As in Proposition 3.12, we get

$$F_1'/F_2' \cong H^1(\Lambda, H^1(\Gamma, \mathbb{Z})) .$$

It follows that $\mathrm{rk}(F_1'/F_2') = 4$. Since $b_2(X) = 4$ we obtain that $F_2'$ is a torsion subgroup of $H^2(G, \mathbb{Z}) \cong H^2(X, \mathbb{Z})$ and that $F_1' = F_0' = H^2(G, \mathbb{Z})$ (by (3.46) $F_0'/F_1'$ is included in $H^0(\Lambda, H^2(\Gamma, \mathbb{Z})) \cong \mathbb{Z}$ , so it is torsion-free, hence it is zero). It follows that $F_1' = F_0' \cong F_0 = F_1$ , where

$$0 \subset F_2 \subset F_1 \subset F_0 = H^2(X, \mathbb{Z})$$

is the filtration induced by the Leray spectral sequence.

Similarly to the remark which follows after the Proposition 3.12, we get:

$$F_2' = \mathrm{Im}(H^2(\Lambda, \mathbb{Z}) \overset{\inf}{\to} H^2(G, \mathbb{Z})) \cong F_2 = \mathrm{Tors}\, H^2(X, \mathbb{Z}).$$

By [ML], Chapter XI, 10, we know that $F_1'$ is the kernel of the restriction. Since $F_1' = H^2(G, \mathbb{Z})$ it follows that this map is zero.

We have:

**Lemma 3.25** *The canonical surjection* $NS(X) \to NS(X)/\mathrm{Tors}\, NS(X)$ *can be identified with the restriction of the homomorphism* $v$ *(defined in* (3.42)*) to* $NS(X)$.

*Proof.* By Lemma 3.24 we have

$$\mathrm{Tors}\, NS(X) = \mathrm{Tors}\, H^2(X, \mathbb{Z}) \cong \mathrm{Im}(\inf) = F_2'$$

and

$$NS(X)/\mathrm{Tors}\, NS(X) \hookrightarrow H^1(\Lambda, H^1(\Gamma, \mathbb{Z})) .$$

Since the restriction map (3.36) is zero we can apply Lemma 3.23 and the conclusion follows.

**Lemma 3.26** *Let* $X$ *be a primary Kodaira surface. Then the following group* $NS(X)/\mathrm{Tors}\, NS(X)$ *is isomorphic with the subgroup of* $\mathcal{M}_2(\mathbb{Z})$

$$\mathcal{NS} := \left\{ A = \begin{pmatrix} A & B \\ C & D \end{pmatrix} \in \mathcal{M}_2(\mathbb{Z}) \mid (B\beta_1 - A\beta_2)\alpha_4 = (D\beta_1 - C\beta_2)\alpha_3 \right\} \qquad (3.49)$$

*Proof.* Since $\mathrm{NS}(X)/\mathrm{Tors}\,\mathrm{NS}(X)$ is regarded as a subgroup of the cohomology group $H^1(\Lambda, H^1(\Gamma, \mathbb{Z}))$ , we have to renounce to a "natural identification". More precisely, we have to consider the isomorphism

$$\mathrm{NS}(X)/\mathrm{Tors}\,\mathrm{NS}(X) \cong \mathrm{Hom}(B, E'),$$

where $E'$ is the dual of $E$ ; $E' = \mathrm{Pic}_0(E) = \mathbb{C}'/\Gamma'$ and $\Gamma' = \mathrm{Hom}_{\mathbb{Z}}(\Gamma, \mathbb{Z}) \cong H^1(\Gamma, \mathbb{Z})$ is the dual lattice in the "complex space" $\mathbb{C}' = \mathrm{Hom}_{\mathbb{Z}}(\Gamma, \mathbb{R})$ (see, for example, [Ke], 1.4). Because $\Gamma$ is a lattice in $\mathbb{C}$ we can extend uniquely any $f \in \mathrm{Hom}_{\mathbb{Z}}(\Gamma, \mathbb{R})$ to a real linear map $\overline{f} : \mathbb{C} \to \mathbb{R}$. Thus

$$\mathrm{Hom}_{\mathbb{Z}}(\Gamma, \mathbb{R}) \cong \mathrm{Hom}_{\mathbb{R}}(\mathbb{C}, \mathbb{R}) = \mathbb{C}'$$

and we put a complex structure on this real vector space defining $if(\gamma) = -\overline{f}(i\gamma)$ , $\gamma \in \Gamma$.

Now, by Corollary 3.14, we have the isomorphism

$$\mathrm{NS}(X)/\mathrm{Tors}\,\mathrm{NS}(X) \cong \{h : \mathbb{C} \to \mathbb{C}' \mid h \,\, \mathbb{C} - \text{linear}, h(\Lambda) \subset \Gamma'\}. \tag{3.50}$$

Any such $h : \mathbb{C} \to \mathbb{C}'$ has the form $h(z) = z \cdot a$ (!), where $a = h(1) \in \mathbb{C}' = \mathrm{Hom}_{\mathbb{Z}}(\Gamma, \mathbb{R})$ and, moreover, must verify the conditions

$$h(\alpha_3) = \alpha_3 \cdot a \in \Gamma' , \quad h(\alpha_4) = \alpha_4 \cdot a \in \Gamma' . \tag{3.51}$$

Denote $a(\beta_1) = t_1, a(\beta_2) = t_2$ $(a = t_1\beta_1' + t_2\beta_2', \,\, t_1, t_2 \in \mathbb{R}$, where $\{\beta_1', \beta_2'\}$ is the dual basis of $\{\beta_1, \beta_2\})$ and let $T = t_2\beta_1 - t_1\beta_2$ . The conditions (3.51) are equivalent with the following conditions:

$$\alpha_3 \cdot a(\beta_1) = A, \alpha_3 \cdot a(\beta_2) = B, \alpha_4 \cdot a(\beta_1) = C, \alpha_4 \cdot a(\beta_2) = D \in \mathbb{Z}. \tag{3.52}$$

Then, by direct computation (with the "complex structure" defined above), we obtain the relations

$$(B\beta_1 - A\beta_2)/\alpha_3 = (D\beta_1 - C\beta_2)/\alpha_4 = T. \tag{3.53}$$

Conversely, writing $T = t_2\beta_1 - t_1\beta_2$ with $t_1, t_2 \in \mathbb{R}$ , and defining $a \in \mathbb{C}'$ by $a(\beta_1) = t_1, a(\beta_2) = t_2$ we obtain a $\mathbb{C}$-linear map $h : \mathbb{C} \to \mathbb{C}'$ , with $h(\Lambda) \subset \Gamma'$.

*Remark.* A matrix $\mathcal{A} \in \mathcal{NS}$ will be our first "ingredient" in the determination of the Picard group of a primary Kodaira surface.

Let $X$ be a complex manifold. Recall that the Picard group $\mathrm{Pic}(X)$ of holomorphic line bundles on $X$ is naturally isomorphic to $H^1(X, \mathcal{O}_X^*)$. The exponential sequence of sheaves

$$0 \to \mathbb{Z}_X \to \mathcal{O}_X \overset{exp}{\to} \mathcal{O}_X^* \to 0 \tag{3.54}$$

gives rise to the cohomology sequence

$$\to H^1(X, \mathbb{Z}) \to H^1(X, \mathcal{O}_X) \to H^1(X, \mathcal{O}_X^*) \overset{\delta}{\to} H^2(X, \mathbb{Z}) \to . \tag{3.55}$$

For any $L \in \text{Pic}(X)$, $\delta(L) = c_1(L)$ is the Chern class of the line bundle. Then $\text{Pic}_0(X) = \text{Ker } \delta$. The subgroup $\text{Pic}^\tau(X)$ of $\text{Pic}(X)$ is defined as the kernel of the composition map $i \circ \delta$, where $i : H^2(X, \mathbb{Z}) \to H^2(X, \mathbb{C})$ is the natural morphism. Clearly, $\text{Pic}^\tau(X)$ consists of the holomorphic line bundles with $c_1(L)$ a torsion element in $H^2(X, \mathbb{Z})$.

**Lemma 3.27** *Let $X$ be a complex compact manifold which has the property that $H^1(X, \mathbb{C}) \to H^1(X, \mathcal{O}_X)$ is surjective. Then $\text{Pic}^\tau(X) = \text{Im}(\rho)$, where $\rho$ is the natural morphism $H^1(X, \mathbb{C}^*) \to H^1(X, \mathcal{O}_X^*)$.*

*Proof.* Standard by diagram chasing.

*Remark.* The map $H^1(X, \mathbb{C}) \to H^1(X, \mathcal{O}_X)$ is always surjective in the case of surfaces (see, for example [B-P-V], Chapter IV).

From now, let $X$ be again a primary Kodaira surface. We have the isomorphism

$$H^1(X, \mathbb{C}^*) \cong H^1(G, \mathbb{C}^*) = \text{Hom}_{\mathbb{Z}}(G, \mathbb{C}^*) \; ; \tag{3.56}$$

see [Mm4], Chapter I, 2, Appendix. With this identification we get the following result:

**Lemma 3.28** *Let $X$ be a primary Kodaira surface. Then*

$$\text{Ker } \rho = \{u \in \text{Hom}_{\mathbb{Z}}(G, \mathbb{C}^*) : u(g) = e^{2\pi i a \lambda}, g = \gamma s(\lambda), a \in \mathbb{C}\}.$$

*Proof.* Here we identify $H^1(X, \mathbb{C}^*)$ with $H^1(G, \mathbb{C}^*)$ and $H^1(X, \mathcal{O}_X^*)$ with $H^1(G, H^*)$; see Section 3.2.
If $u \in \text{Ker } \rho$, then $\rho(u) = \text{cls.}(1)$, i.e. $u$ is cohomologous with the trivial 1-cocycle in the group $H^1(G, H^*)$. It follows that we can find $h \in H^* = H^0(\mathbb{C}^2, \mathcal{O}_{\mathbb{C}^2}^*)$ such that

$$h(g(z, w)) = u(g)h(z, w) \; , \; g \in G \; , \; (z, w) \in \mathbb{C}^2. \tag{3.57}$$

For $g = \gamma \in \Gamma$ we get

$$h(z, w + \gamma) = u(\gamma)h(z, w) \; , \; \gamma \in \Gamma \; , \; (z, w) \in \mathbb{C}^2. \tag{3.58}$$

Fixing $z$ and taking logarithmic derivatives with respect to $w$ ( so as to eliminate $u(\gamma)$), we get for the holomorphic function $\omega = h'_w/h$ the relation

$$\omega(w + \gamma) = \omega(w) . \gamma \in \Gamma , w \in \mathbb{C}. \tag{3.59}$$

If $K$ is a compact set in $\mathbb{C}$ with $K + \Gamma = \mathbb{C}$, it follows that for any $w \in \mathbb{C}$,

$$|\omega(w)| \leq \sup_K |\omega(w)|.$$

Hence $\omega$ can only be a constant in $w$, say $b(z)$. We obtain:

$$h(z, w) = e^{2\pi i (b(z)w + c(z))} \; , \; (z, w) \in \mathbb{C}^2 \; , \tag{3.60}$$

with $b(z), c(z)$ holomorphic functions on $\mathbb{C}$. But from (3.58) we get $b(z) = b$ a constant. By (3.57) and (3.60) we obtain for $g = s(\lambda) \in G$

$$u(s(\lambda)) = e^{2\pi i(b(\bar{\lambda}z + \beta(s(\lambda))) + c(z+\lambda) - c(z))}.$$

Taking the derivatives with respect to $z$ and exponentiating we get

$$e^{2\pi i c'(z+\lambda)} = e^{-2\pi i b\bar{\lambda}} \cdot e^{2\pi i c'(z)}.$$

Taking again logarithmic derivatives we obtain as above $c'(z) = dz + a$. It follows $d\lambda + b\bar{\lambda} = 0$ for all $\lambda \in \Lambda$, with $b, d \in \mathbb{C}$, which implies $d = b = 0$. Finally, we get

$$h(z, w) = e^{2\pi i(az+k)}, a, k \in \mathbb{C}, \tag{3.61}$$

and, therefore, by (3.57)

$$u(g) = e^{2\pi i a\lambda}, g = \gamma s(\lambda) \in G.$$

The converse is obvious.

**Lemma 3.29** *The group $Pic^\tau(X)$ is isomorphic with the following subgroup of $Hom_\mathbb{Z}(G, \mathbb{C}^*)$*

$$\mathcal{P} = \{u \in Hom_\mathbb{Z}(G, \mathbb{C}^*) : u(s(\lambda)) \in U(1) , \lambda \in \Lambda\}.$$

*Proof.* From the relation $g_3 g_4 = g_1^m g_4 g_3$ we get for any $u \in Hom_\mathbb{Z}(G, \mathbb{C}^*)$ that $u(g_1^m) = 1$ and this implies

$$u(s(\lambda))u(s(\lambda')) = u(s(\lambda + \lambda')) , \lambda, \lambda' \in \Lambda.$$

Writing $u(s(\lambda)) = e^{2\pi i r(\lambda)}$ it follows that

$$r(\lambda) + r(\lambda') - r(\lambda + \lambda') \in \mathbb{Z} , \lambda, \lambda' \in \Lambda, \tag{3.62}$$

which implies $\operatorname{Im} r(\lambda) + \operatorname{Im} r(\lambda') = \operatorname{Im} r(\lambda + \lambda')$. The $\mathbb{Z}$-linear map $\varphi : \Lambda \to \mathbb{R}$ , $\varphi(\lambda) = \operatorname{Im} r(\lambda)$ has a unique real extension $\bar{\varphi} : \mathbb{C} \to \mathbb{R}$. Take the complex linear mapping $k : \mathbb{C} \to \mathbb{C}$, $k(z) = \operatorname{Im} \bar{\varphi}(iz) + i \operatorname{Im} \bar{\varphi}(z)$ ( see, for example, [Ke], p. 6). Then $\bar{r} = \bar{\varphi} - k$ is real valued. Writing $k(z) = az$ for some $a \in \mathbb{C}$ we take $u_0 \in \operatorname{Ker} \rho$ with $u_0(g) = e^{2\pi i a\lambda}$ (Lemma 3.28). The quotient $u/u_0 = \bar{u}$ has, clearly, the property $\bar{u}(s(\lambda)) \in U(1)$ and it is uniquely determined by this property in the class of $u$ ($\rho(u) = \operatorname{cls.}(u) \in Pic^\tau(X)$).

Let $X$ be a primary Kodaira surface and let $\mathbb{C}^2 \to X$ be its universal covering. Recall that we denoted by $H^*$ the multiplicative group $H^0(\mathbb{C}^2, \mathcal{O}_{\mathbb{C}^2}^*)$ of non-vanishing holomorphic functions on $\mathbb{C}^2$. If $H$ is the ring of holomorphic functions on $\mathbb{C}^2$, we have the exact sequence

$$0 \to \mathbb{Z} \to H \xrightarrow{exp} H^* \to 0, \tag{3.63}$$

and by [Mm4], Chapter I, 2, Appendix, we get the diagram

$$H^1(G, H^*) \xrightarrow{\delta} H^2(G, \mathbb{Z})$$

$$\wr \downarrow \qquad\qquad \wr \downarrow$$

$$H^1(X, \mathcal{O}_X^*) \xrightarrow{\delta} H^2(X, \mathbb{Z})$$

Recall that a line bundle $L$ on $X$ is given by the class of a cocycle $(e_g) \in H^1(G, H^*)$. By the above identifications it follows that the Chern class of $L$ is $\delta(\text{cls.}(e_g))$. If we write

$$e_g(z, w) = e^{2\pi i f_g(z,w)}, \tag{3.64}$$

with $f_g$ holomorphic in $\mathbb{C}^2$ then, by definition, $\delta(\text{cls.}(e_g)) \in H^2(G, \mathbb{Z})$ is given by the 2-cocycle $F(g, g')$ on $G$ with coefficients in $\mathbb{Z}$ given by

$$F(g, g') = f_g(g'(z, w)) - f_{gg'}(z, w) + f_{g'}(z, w) \in \mathbb{Z}. \tag{3.65}$$

Now, our problem is equivalent to finding a system of "simple" functions $(f_g)_{g \in G}$ holomorphic in $\mathbb{C}^2$ and satisfying (3.65).

We shall look for solutions $f_g$ which are linear in $(z, w)$

$$f_g(z, w) = p(g)z + q(g)w + r(g), \tag{3.66}$$

where $p, q, r : G \to \mathbb{C}$.

From the condition (3.65) we get the relations:

$$p(gg') = p(g) + p(g') + \overline{\lambda} q(g'), \tag{3.67}$$

$$q(gg') = q(g) + q(g'), \tag{3.68}$$

and

$$F(g, g') = p(g)\lambda' + q(g)\beta(g') + r(g) + r(g') - r(gg') \in \mathbb{Z}, \tag{3.69}$$

where $g = \gamma s(\lambda), g' = \gamma' s(\lambda')$, all $g, g' \in G$.

By (3.68) $q$ is a homomorphism, so it has the form

$$q(g) = q_2 l_2 + q_3 l_3 + q_4 l_4, \ g \in G.$$

Since $g_2 g_3 = g_3 g_2$, it follows by (3.67) that $q_2 = 0$, i.e. $q$ is a linear form depending only on $\lambda \in \Lambda$:

$$q(g) = q_3 l_3 + q_4 l_4 = q(s(\lambda)), g \in G. \tag{3.70}$$

From (3.67) we get that $p(g)$ is a quadratic polynomial of the form

$$p(g) = p_1 l_1 + p_2 l_2 + \tilde{p}_3 l_3 + \tilde{p}_4 l_4 + (1/2)\overline{\alpha}_3 q_3 l_3^2 + (1/2)\overline{\alpha}_4 q_4 l_4^2 + \overline{\alpha}_3 q_4 l_3 l_4. \tag{3.71}$$

Using the relation

$$g_3 g_4 = g_1^m g_4 g_3$$

we obtain the equation

$$\overline{\alpha}_4 q_3 - \overline{\alpha}_3 q_4 = m p_1. \tag{3.72}$$

It will become clear later that we can take $\tilde{p}_3 = \tilde{p}_4 = 0$; then, with a notation similar to that of (3.31), we have

$$p(g) = p(\gamma) + p_2(s(\lambda)), \ g = \gamma s(\lambda) \in G, \tag{3.73}$$

where $p(\gamma)$ is a linear form depending only on $\gamma \in \Gamma$ and $p_2(s(\lambda))$ is a quadratic form depending only on $\lambda \in \Lambda$.

Finally, we obtain for $F(g, g')$ the form

$$F(g, g') = (p(\gamma)\lambda' + q(s(\lambda))\gamma') + (q(s(\lambda))\beta_1(s(\lambda'))) + \tag{3.74}$$

$$+ (p_2(s(\lambda))\lambda' + q(s(\lambda))\beta_2(s(\lambda'))) + (r(g) + r(g') - r(gg')) \in \mathbb{Z}.$$

Now we shall simplify in several steps the form of the $F(g, g')$ by modifying $r(g)$.

*Step 1.* We shall transform the quadratic form $q(s(\lambda))\beta_1(s(\lambda'))$ in a simpler one. Put

$$r(g) = r_1(g) + (1/2)(q(s(\lambda))\beta_1(s(\lambda))). \tag{3.75}$$

Computing we get

$$F(g, g') = (p(\gamma)\lambda' + q(s(\lambda))\gamma') + t(l_3 l_4' - l_3' l_4) + \tag{3.76}$$

$$+ (p_2(s(\lambda))\lambda' + q(s(\lambda))\beta_2(s(\lambda'))) + (r_1(g) + r_1(g') - r_1(gg')) \in \mathbb{Z},$$

where $t = (1/2)(q_3\tilde{\beta}_4 - q_4\tilde{\beta}_3)$.

*Step 2.* We shall "hide" the quadratic form $t(l_3 l_4' - l_3' l_4)$. Let $h : G \to \mathbb{C}$ be the map $h(g) = (t/m)l_1$, all $g \in G$, and define $\tilde{h} : G \to \mathbb{C}$ by

$$\tilde{h}(g) = h(g^{-1}) = (t/m)(-l_1 + ml_3l_4).$$

Put

$$r_1(g) = r_2(g) + h(g) - \tilde{h}(g). \tag{3.77}$$

Computing we get

$$F(g, g') = (p(\gamma)\lambda' + q(s(\lambda))\gamma') + (p_2(s(\lambda))\lambda' + q(s(\lambda))\beta_2(s(\lambda'))) + \tag{3.78}$$

$$+ (r_2(g) + r_2(g') - r_2(gg')) \in \mathbb{Z}.$$

We see now that we could choose $\tilde{p}_3 = \tilde{p}_4 = 0$ (because the contribution of these terms to $F(g, g')$ is "hidden" in $r_2(g)$).

*Step 3.* We shall simplify the cubic form of (3.78) by

$$r_2(g) = r_3(g) + (al_3^3 + bl_3^2 l_4 + cl_3 l_4^2 + dl_4^3), \tag{3.79}$$

imposing the vanishing of the coefficients of all terms, but those which contain the product $l_3 l_4'$. We get

$$a = (1/6)\overline{\alpha}_3\alpha_3 q_3 \ , \ d = (1/6)\overline{\alpha}_4\alpha_4 q_4, \tag{3.80}$$

$$b = (1/2)\overline{\alpha}_3\alpha_3 q_4 \ , \ c = (1/2)\overline{\alpha}_4\alpha_3 q_4.$$

Finally, we have

$$F(g, g') = (p(\gamma)\lambda' + q(s(\lambda))\gamma') + \tag{3.81}$$

$$+(1/2)l_3l_4'((\overline{\alpha}_3\alpha_4q_3 - \overline{\alpha}_3\alpha_3q_4)l_3 + (\overline{\alpha}_4\alpha_4q_3 - \overline{\alpha}_4\alpha_3q_4)l_4')+$$

$$+l_3l_4'((\overline{\alpha}_3\alpha_4q_4 - \overline{\alpha}_4\alpha_3q_4)l_4 + (\overline{\alpha}_4\alpha_3q_3 - \overline{\alpha}_3\alpha_3q_4)l_3')+$$

$$+(r_3(g) + r_3(g') - r_3(gg')) \in \mathbb{Z}.$$

By using the formula (3.81) and by applying Lemma 3.22 we shall compute a 2-cocycle $\tilde{F}$ associated with $F$ (and cohomologous to it). We know that the restriction of $F$ to $\Gamma \times \Gamma$ is cohomologous to zero in $H^2(\Gamma, \mathbb{Z})$ ( Lemma 3.24) and we need the explicit computation from the proof of Lemma 3.22. We have

$$F(\gamma, \gamma') = r_3(\gamma) + r_3(\gamma') - r_3(\gamma + \gamma') \in \mathbb{Z}, \ \gamma, \gamma' \in \Gamma. \tag{3.82}$$

We observe that this cocycle is a coboundary in $H^2(\Gamma, \mathbb{C})$ and, because $H^2(\Gamma, \mathbb{Z})$ has no torsion, we get that this cocycle is a coboundary in $H^2(\Gamma, \mathbb{Z})$ too. Anyway, there exists a map $\theta : \Gamma \to \mathbb{Z}$ such that

$$F(\gamma, \gamma') = (\delta\theta)(\gamma, \gamma'), \gamma, \gamma' \in \Gamma. \tag{3.83}$$

Define the function $\Omega : G \to \mathbb{Z}$ by

$$\Omega(g) = F(\gamma, s(\lambda)) - \theta(\gamma), g = \gamma s(\lambda) \in G. \tag{3.84}$$

By (3.81) we obtain

$$\Omega(g) = p(\gamma)\lambda + r_3(\gamma) + r_3(s(\lambda)) - r_3(g) - \theta(\gamma) \in \mathbb{Z}. \tag{3.85}$$

From Lemma 3.22 we know that

$$\tilde{F}(g, g') = F(g, g') + \Omega(g) + \Omega(g') - \Omega(gg').$$

Using (3.81), (3.83) and (3.85) we get the formula:

$$\tilde{F}(g, g') = (q(s(\lambda))\gamma - p(\gamma')\lambda)+ \tag{3.86}$$

$$+l_3l_4'(((1/2)(\overline{\alpha}_3\alpha_4q_3 - \overline{\alpha}_3\alpha_3q_4) - m\alpha_3p_1)l_3+$$

$$+((1/2)(\overline{\alpha}_4\alpha_4q_3 - \overline{\alpha}_4\alpha_3q_4) - m\alpha_4p_1)l_4')+$$

$$+l_3l_4'((\overline{\alpha}_4\alpha_3q_3 - \overline{\alpha}_3\alpha_3q_4 - m\alpha_3p_1)l_3' + (\overline{\alpha}_3\alpha_4q_4 - \overline{\alpha}_4\alpha_3q_4 - m\alpha_4p_1)l_4)+$$

$$+(r_3(s(\lambda)) + r_3(s(\lambda')) - r_3(s(\lambda + \lambda')) - r_3(g_1^{ml_3l_4'}) + \theta(g_1^{ml_3l_4'})) \in \mathbb{Z}.$$

Now apply Lemma 3.23 and compute:

$$\overline{F}(\lambda, \gamma') = q(s(\lambda))\gamma' - p(\gamma')\lambda = \tag{3.87}$$

$$= (\beta_1q_3 - \alpha_3p_1)l_1'l_3 + (\beta_2q_3 - \alpha_3p_2)l_2'l_3+$$

$$+(\beta_1q_4 - \alpha_4p_1)l_1'l_4 + (\beta_2q_4 - \alpha_4p_2)l_2'l_4 \in \mathbb{Z}$$

(we used also the fact that all cocycles are normalized).

By Lemma 3.25 it follows that $\overline{F}(\lambda, \gamma')$ should be the Chern class of the line bundle $L$, modulo torsion. Then $A = \beta_1q_3 - \alpha_3p_1$, $B = \beta_2q_3 - \alpha_3p_2$, $C = \beta_1q_4 - \alpha_4p_1$, $D = \beta_2q_4 - \alpha_4p_2$ must be integers.

Let's point out the following trivial, but useful fact:

**Lemma 3.30** *The system of linear equations for* $p_1, p_2, q_3, q_4$

$$\beta_1 q_3 - \alpha_3 p_1 = A, \beta_2 q_3 - \alpha_3 p_2 = B, \beta_1 q_4 - \alpha_4 p_1 = C, \tag{3.88}$$

$$\beta_2 q_4 - \alpha_4 p_2 = D, \overline{\alpha}_4 q_3 - \overline{\alpha}_3 q_4 - m p_1 = 0$$

*has a unique solution iff* $(B\beta_1 - A\beta_2)\alpha_4 = (D\beta_1 - C\beta_2)\alpha_3$.

Thus, if we have a Chern class modulo torsion $A \in \mathcal{NS}$ (see Lemma 3.26) then $q(g)$ and $p(g)$ are uniquely determined by $A$ (see (3.70) and (3.73)).

By using (3.88) the formula (3.86) becomes:

$$\tilde{F}(g, g') = (Al'_1 l_3 + Bl'_2 l_3 + Cl'_1 l_4 + Dl'_2 l_4) + \tag{3.89}$$

$$+ m l_3 l'_4 (Al_3 + Cl'_4 + 2Cl_4)/2 +$$

$$+ (r_3(s(\lambda)) + r_3(s(\lambda')) - r_3(s(\lambda + \lambda')) - r_3(g_1^{m l_3 l'_4}) + \theta(g_1^{m l_3 l'_4})) \in \mathbb{Z}.$$

It follows from (3.88) that $r_3(s(\lambda))$ must verify the condition

$$r_3(s(\lambda)) + r_3(s(\lambda')) - r_3(s(\lambda + \lambda')) - r_3(g_1^{m l_3 l'_4}) + \tag{3.90}$$

$$+ m l_3 l'_4 (Al_3 + Cl'_4)/2 \in \mathbb{Z}.$$

Taking in (3.90) $\lambda = \alpha_3$, $\lambda' = \alpha_4$, then $\lambda = \alpha_4$, $\lambda' = \alpha_3$ and substracting the results, we get

$$r_3(g_1^m) - m(A + C)/2 \in \mathbb{Z}. \tag{3.91}$$

By (3.82) we have

$$r_3(g_1^{m l_3 l'_4}) - m l_3 l'_4 (A + C)/2 \in \mathbb{Z},$$

hence (3.90) becomes

$$r_3(s(\lambda)) + r_3(s(\lambda')) - r_3(s(\lambda + \lambda')) + \tag{3.92}$$

$$+ m l_3 l'_4 (A(l_3 - 1) + C(l'_4 - 1))/2 \in \mathbb{Z}.$$

Since

$$l_3 l'_4 (A(l_3 - 1) + C(l'_4 - 1)) \equiv 0 \pmod 2,$$

for any $l_3, l'_4 \in \mathbb{Z}$, it follows that $r_3(s(\lambda))$ must satisfy the condition:

$$r_3(s(\lambda)) + r_3(s(\lambda')) - r_3(s(\lambda + \lambda')) \in \mathbb{Z} \tag{3.93}$$

Writing

$$u(g) = e^{2\pi i r_3(g)}$$

for all $g \in G$, we see, by (3.83), (3.85), (3.91) and (3.93), that $u$ has to satisfy the conditions:

$$\begin{cases} u(g_1^m) & = & e^{\pi i m(A+C)}, \\[2mm] u(\gamma + \gamma') & = & u(\gamma)u(\gamma'), \gamma, \gamma' \in \Gamma, \\[2mm] u(s(\lambda + \lambda')) & = & u(s(\lambda))u(s(\lambda')), \lambda, \lambda' \in \Lambda, \\[2mm] u(g) & = & u(\gamma)u(s(\lambda))e^{2\pi i p(\gamma)\lambda}, \end{cases} \tag{3.94}$$

where $p(\gamma)$ is determined by (3.88).

Modifying the cocycle $f_g$ by a coboundary, with the same method as in the proof of Lemma 3.29 (using a suitable function of the form $h(z) = e^{2\pi i a z}$), we may assume that $r_3(s(\lambda))$ is real. Then:

$$u(s(\lambda)) \in U(1). \tag{3.95}$$

**Definition 3.31** *Appell-Humbert data* are a pair $(u, \mathcal{A})$ where $\mathcal{A} \in \mathcal{NS}$ (the subgroup of $\mathcal{M}_2(\mathbb{Z})$ from Lemma 3.26) and $u$ is a mapping from $G$ to $\mathbb{C}^*$ satisfying the conditions (3.94) and (3.95).

*Remark.* Obviously, all the Appel-Humbert data $(u, \mathcal{A})$ form a group with the law

$$(u, \mathcal{A}) \cdot (u', \mathcal{A}') = (uu', \mathcal{A} + \mathcal{A}').$$

For a given pair $(u, \mathcal{A})$ we determine $q(g), p(g)$ as above and, by (3.79), (3.77) and (3.75), we determine $r(g)$, hence the cocycle $(e_g)$. Denoting by $L(u, \mathcal{A})$ the line bundle on $X$ defined by the cocycle $(e_g)$ we get a map

$$\Psi : \{\text{Group of data } (u, \mathcal{A})\} \to \text{Pic}(X), \Psi(u, \mathcal{A}) = L(u, \mathcal{A}),$$

which is, clearly, a homomorphism.

**Theorem 3.32** *Any line bundle $L$ on the primary Kodaira surface $X$ is isomorphic to a line bundle $L(u, \mathcal{A})$ for a uniquely determined pair $(u, \mathcal{A})$. Moreover, we have the following isomorphic exact sequences:*

$$
\begin{array}{ccccccccc}
0 & \longrightarrow & \mathcal{P} & \longrightarrow & \{\text{Group of data } (u, \mathcal{A})\} & \longrightarrow & \mathcal{NS} & \longrightarrow & 0 \\
& & \wr \downarrow \Psi' & & \wr \downarrow \Psi & & \wr \downarrow \Psi'' & & \\
0 & \longrightarrow & \text{Pic}^\tau(X) & \longrightarrow & \text{Pic}(X) & \longrightarrow & \text{NS}(X)/\text{Tors NS}(X) & \longrightarrow & 0
\end{array}
$$

*where $\Psi'$ is defined in Lemma 3.29 and $\Psi''$ is defined in Lemma 3.26.*

*Proof.* We see that for any $\mathcal{A} \in \mathcal{NS}$ we can find $u$ verifying (3.94) and (3.95); $p(\gamma)$ is determined and we can define $u$ on generators $\{g_1, g_2\}$, respectively $\{g_3, g_4\}$. Then, if $\mathcal{A}$ is zero, by (3.94) it follows that $u$ is a homomorphism and by (3.95) we get $u \in \mathcal{P}$. Thus the first sequence is exact. By the proofs of the Lemmas 3.26 and 3.29 it is clear that the diagrams commute and, since $\Psi'$ and $\Psi''$ are isomorphisms, $\Psi$ is an isomorphism.

# 4. Existence of holomorphic vector bundles

In this chapter we present (partial) solutions to the existence problem for holomorphic vector bundles over compact complex surfaces. We describe Serre method of constructing holomorphic vector bundles over complex manifolds as extensions and we give a proof of the existence result of Schwarzenberger for the case of algebraic surfaces. In the case of nonalgebraic surfaces we present the complete answer to the existence of filtrable holomorphic vector bundles and a necessary condition for (general) holomorphic structures (see Bănică-Le Potier [B-L]). Then we give examples of non-filtrable and even irreducible holomorphic vector bundles over nonalgebraic surfaces, whose presence makes difficult the existence problem, still open in this case. Finally, we give a complete answer to the existence of simple filtrable holomorphic 2-vector bundles over surfaces with algebraic dimension zero.

## 4.1 Serre construction

The main method of constructing holomorphic vector bundles is the Serre construction, which we present following essentially [O-S-S], Chapter I, 5.1.

Let $X$ be a compact complex manifold. Let $E$ be a holomorphic vector bundle of rank 2 over $X$ (shortly, 2-vector bundle) with a non-zero section $s \in H^0(X, E)$. Let $D$ be the zero divisor associated with $s$. Locally, we can describe the situation as follows. Let $U \subset X$ be an open set such that $E|_U$ is trivial and let $s_1$, $s_2 \in H^0(U, E|_U)$ be a local basis (local frame) for $E$ over $U$. Then

$$s|_U = f_1 s_1 + f_2 s_2$$

for appropriate holomorphic functions $f_1, f_2 \in H^0(U, \mathcal{O})$. Let $f_1 = \theta \tilde{f}_1$, $f_2 = \theta \tilde{f}_2$ with $\theta$ a holomorphic function on a (possible) smaller open set $\tilde{U} \subset U$ such that on $\tilde{U}$ the functions $\tilde{f}_1$, $\tilde{f}_2$ have a common zero set of codimension 2 or the empty set. Then, the local equation of the zero divisor $D$ is exactly the holomorphic function $\theta$. Consider the holomorphic 2-vector bundle $E \otimes \mathcal{O}_X(-D)$. Then we get a non-zero section $\tilde{s} \in H^0(X, E \otimes \mathcal{O}_X(-D))$ such that, locally on an open set $\tilde{U}$, we have

$$\tilde{s}|_{\tilde{U}} = \tilde{f}_1 s_1 + \tilde{f}_2 s_2$$

and the functions $\tilde{f}_1$, $\tilde{f}_2$ have a common zero set of codimension 2 or the empty set. Now, we shall replace the holomorphic 2-vector bundle $E$ by the holomorphic 2-vector bundle $\tilde{E} = E \otimes \mathcal{O}_X(-D)$ and we suppose that there exists a non-zero section

$\tilde{s} \in H^0(X, \tilde{E})$ whose zero set $Y$ is of codimension 2 in $X$. For the global sheaf of ideals $J_Y \subset \mathcal{O}_X$ we have

$$J_Y|_{\tilde{U}} = (\tilde{f}_1, \tilde{f}_2)\mathcal{O}_X|_{\tilde{U}} .$$

Obviously

$$\text{Supp}(\mathcal{O}_X/J_Y) = Y$$

and $(Y, \mathcal{O}_X/J_Y)$ is a codimension 2 locally complete intersection in $X$, denoted simply $Y$ and called the *zero locus* of the section $\tilde{s} \in H^0(X, \tilde{E})$. Of course, $Y$ can be non-reduced.

From the local situation described above we see that the germs $\tilde{f}_{1,x}$, $\tilde{f}_{2,x}$ for any $x \in \tilde{U} \cap Y$ form a regular sequence and represent an $\mathcal{O}_{Y,x}$-module basis of $J_{Y,x}/J_{Y,x}^2$. It follows that the $\mathcal{O}_Y$-module $J_Y/J_Y^2$ is locally free of rank 2. It is the conormal bundle of $Y$ in $X$ and its dual $N_{Y/X} = (J_Y/J_Y^2)^*$ is the normal bundle of $Y$ in $X$ (see also Section 1.1). Since the germs $\tilde{f}_{1,x}$, $\tilde{f}_{2,x}$ form a regular sequence, it follows that, locally, the sheaf $J_Y$ has a free resolution

$$0 \to \mathcal{O}_{\tilde{U}} \xrightarrow{\alpha} \mathcal{O}_{\tilde{U}} \oplus \mathcal{O}_{\tilde{U}} \xrightarrow{\beta} J_Y|_{\tilde{U}} \to 0 ,$$

where

$$\alpha(g) = (-\tilde{f}_{2,x}g, \tilde{f}_{1,x}g)$$

and

$$\beta(g, h) = \tilde{f}_{1,x}g + \tilde{f}_{2,x}h .$$

Globally, we get a resolution

$$0 \to \det(\tilde{E}^*) \xrightarrow{\alpha} \tilde{E}^* \xrightarrow{\beta} J_Y \to 0 ,$$

where $\tilde{E}^* = \mathcal{H}om(\tilde{E}, \mathcal{O}_X)$ is the dual vector bundle of $\tilde{E}$,

$$\alpha(\varphi_1 \wedge \varphi_2) = \varphi_1(\tilde{s}_x)\varphi_2 - \varphi_2(\tilde{s}_x)\varphi_1 , \quad \beta(\varphi) = \varphi(\tilde{s}_x)$$

for $x \in X$ and

$$\varphi_1, \varphi_2, \varphi \in \tilde{E}_x^* = \text{Hom}(\tilde{E}_x, \mathcal{O}_{X,x}) .$$

Combining with the exact sequence

$$0 \to J_Y \to \mathcal{O}_X \to \mathcal{O}_Y \to 0$$

we obtain the *Koszul complex* for the section $\tilde{s}$

$$0 \to \det(\tilde{E}^*) \to \tilde{E}^* \xrightarrow{s^*} \mathcal{O}_X \to \mathcal{O}_Y \to 0 .$$

Tensoring by $\det(\tilde{E})$, since

$$\tilde{E} \cong \tilde{E}^* \otimes \det(\tilde{E}) ,$$

we have the exact sequence

$$0 \to \mathcal{O}_X \to \tilde{E} \to J_Y \otimes \det(\tilde{E}) \to 0 .$$

For the initial 2-vector bundle $E$ we get

$$0 \to L_1 \to E \to J_Y \otimes L_2 \to 0 , \tag{4.1}$$

where $L_1 \cong \mathcal{O}_X(D)$ and $L_2 \cong \det(\tilde{E}) \otimes \mathcal{O}_X(D)$ are line bundles.

We proved the following result:

**Proposition 4.1** *Let $X$ be a compact complex manifold and let $E$ be a holomorphic 2-vector bundle with a non-zero section $s \in H^0(X, E)$. Then there exists an extension*

$$0 \to L_1 \to E \to J_Y \otimes L_2 \to 0,$$

*with $L_1$, $L_2$ line bundles and where $J_Y$ is the ideal sheaf of a codimension 2 locally complete intersection $Y$ in $X$ or $Y$ is empty.*

The Serre method consists in reversing this construction. Starting with $L_1$, $L_2$ in $\mathrm{Pic}(X)$ and $Y$ a locally complete intersection 2-codimensional analytic subspace of $X$, one may ask when there exist extensions of $J_Y \otimes L_2$ through $L_1$

$$0 \to L_1 \to E \to J_Y \otimes L_2 \to 0,$$

such that $E$ be locally free.

The extensions (4.1) are classified by the vector space (global Ext-group)

$$\mathrm{Ext}^1(J_Y \otimes L_2, L_1) \cong \mathrm{Ext}^1(J_Y, L) ,$$

where $L = L_2^* \otimes L_1$. We need criteria to decide that the central term of (4.1) is locally free, i.e. all the fibres of $E$ are free modules. We have the following result of Serre (see [Sr4], [O-S-S], Lemma 5.1.2):

**Lemma 4.2** *Let $A$ be a noetherian local ring, $I \subset A$ an ideal with a free resolution of length 1*

$$0 \to A^p \to A^q \xrightarrow{\varphi} I \to 0 .$$

*Let $e \in \mathrm{Ext}_A^1(I, A)$ be represented by the extension*

$$0 \to A \xrightarrow{\alpha} M \xrightarrow{\beta} I \to 0 .$$

*Then $M$ is a free $A$-module if and only if $e$ generates the $A$-module $\mathrm{Ext}_A^1(I, A)$.*

*Proof.* The exact sequence

$$0 \to A \to M \to I \to 0$$

gives rise to the exact sequence of Ext's:

$$\cdots \to \mathrm{Hom}_A(A, A) \xrightarrow{\delta} \mathrm{Ext}_A^1(I, A) \to \mathrm{Ext}_A^1(M, A) \to 0 .$$

Because $\delta(id_A) = e$ (see [ML], Chapter XII, 5), then $e$ generates the $A$-module $\mathrm{Ext}_A^1(I, A)$ if and only if $\delta$ is surjective, i.e. if and only if $\mathrm{Ext}_A^1(M, A) = 0$.

To end the proof it suffices to show that $\text{Ext}^1_A(M, A) = 0$ implies $M$ is free. Let $\Phi : A^q \to M$ be a lifting of the map $\varphi$ to $M$ such that $\beta \circ \Phi = \varphi$ and define $\Psi : A \oplus A^q \to M$ by $\Psi(x, y) = \alpha(x) + \Phi(y)$. By standard chasing, it follows that $\text{Ker}(\Psi) \cong \text{Ker}(\Phi) \cong A^q$ and $\text{Coker}(\Psi) = 0$. We get an exact sequence:

$$0 \to A^p \to A \oplus A^q \to M \to 0 .$$

Since $\text{Ext}^1_A(M, A) = 0$, this sequence splits. It follows that $M$ is a direct summand in $A^{q+1}$, hence projective and free.

Consider again the vector space of extensions $\text{Ext}^1(J_Y, L)$. By removability (second Riemann extension Theorem) we have

$$\mathcal{H}om(J_Y, L) \cong L ,$$

since $\text{codim}_X Y = 2$. The exact lower terms sequence of the Ext spectral sequence becomes:

$$0 \to H^1(X, L) \to \text{Ext}^1(J_Y, L) \to H^0(X, \mathcal{E}xt^1(J_Y, L)) \to H^2(X, L) .$$

By the above lemma it follows that the extension corresponding to an element $\xi \in \text{Ext}^1(J_Y, L)$ will give a holomorphic 2-vector bundle if and only if the image of $\xi$, through the canonical mapping

$$\text{Ext}^1(J_Y, L) \longrightarrow H^0(X, \mathcal{E}xt^1(J_Y, L))$$

generates the sheaf $\mathcal{E}xt^1(J_Y, L)$.

From the exact $\mathcal{E}xt$-sequence associated to

$$0 \to J_Y \to \mathcal{O}_X \to \mathcal{O}_Y \to 0$$

we get the isomorphism:

$$\mathcal{E}xt^1(J_Y, L) \cong \mathcal{E}xt^2(\mathcal{O}_Y, L) .$$

Since $Y$ is a codimension 2 locally complete intersection, we have the local fundamental isomorphism (see [G-H], p. 690)

$$\mathcal{E}xt^2(\mathcal{O}_Y, L) \cong \mathcal{H}om_{\mathcal{O}_Y}(\det(J_Y/J_Y^2), L|_Y) .$$

Let us suppose that $\det(N_{Y/X}) \cong L^*|_Y$. Then it follows

$$\mathcal{E}xt^2(\mathcal{O}_Y, L) \cong \det(N_{Y/X}) \otimes L|_Y \cong \mathcal{O}_Y ,$$

hence we get the exact sequence

$$\text{Ext}^1(J_Y, L) \to H^0(Y, \mathcal{O}_Y) \to H^2(X, L) .$$

Now, if $H^2(X, L)$ is zero, since $1 \in H^0(Y, \mathcal{O}_Y)$ generates the $\mathcal{O}_{X,x}$-module $\mathcal{O}_{Y,x} \cong \mathcal{E}xt^1(J_Y, L)_x$ for every $x \in X$, then any element $\xi \in \text{Ext}^1(J_Y, L)$ with image $1 \in H^0(Y, \mathcal{O}_Y)$ will give a holomorphic 2-vector bundle.

Thus we proved the following result (see [O-S-S], Chapter I, Theorem 5.1.1):

**Theorem 4.3** *Let $X$ be a compact complex manifold, $L_1$, $L_2$ in $Pic(X)$, and let $Y$ be a locally complete intersection of codimension 2 in $X$. Denote $L = L_2^* \otimes L_1$; suppose that $H^2(X, L) = 0$ and $\det(N_{Y/X}) \cong L^*|_Y$. Then there exists a holomorphic 2-vector bundle $E$ on $X$ such that*

$$0 \to L_1 \to E \to J_Y \otimes L_2 \to 0 .$$

*Remark.* Let $X$ be a compact complex surface. Then:

(1) The condition $\det(N_{Y/X}) \cong L^*|_Y$ is always fulfilled since $\det(N_{Y/X})$ is trivial ($\dim Y = 0$ !).

(2) In applications the condition $H^2(X, L) = 0$ is too strong. We shall use sometimes the result in the following form: we find an element $\xi \in \text{Ext}^2(\mathcal{O}_Y, L)$ which generates the sheaf $\mathcal{E}xt^2(\mathcal{O}_Y, L) \cong \mathcal{O}_Y$ and which is mapped to zero in $H^2(X, L)$. Note also that, by Serre duality, the map

$$\text{Ext}^2(\mathcal{O}_Y, L) \longrightarrow H^2(X, L)$$

is the dual of the restriction map

$$H^0(X, L^* \otimes \mathcal{K}_X) \longrightarrow H^0(Y, L^* \otimes \mathcal{K}_X|_Y) .$$

From the exact sequence

$$0 \to L_1 \to E \to J_Y \otimes L_2 \to 0$$

we get the exact sequence

$$0 \to \mathcal{O}_X \to E \otimes L_1^* \to J_Y \otimes L_2 \otimes L_1^* \to 0 .$$

The morphism $\mathcal{O}_X \to E \otimes L_1^*$ corresponds to the multiplication with a section $s \in H^0(X, E \otimes L_1^*)$, which has precisely $(Y, \mathcal{O}_Y)$ as zero set.

**Proposition 4.4** *Let $X$ be a compact complex manifold and let $E$ be a holomorphic 2-vector bundle over $X$ given by an extension of the form*

$$0 \to L_1 \to E \to J_Y \otimes L_2 \to 0 ,$$

*where $L_1$, $L_2$ are line bundles and $Y$ is a locally complete intersection of codimension 2 in $X$. Then we have:*

$$c_1(E) = c_1(L_1) + c_1(L_2) ,$$
$$c_2(E) = c_1(L_1).c_1(L_2) + [Y] .$$

*Proof.* Restricting the above extension to the open set $U = X \setminus Y$ we get:

$$0 \to L_1|_U \to E|_U \to L_2|_U \to 0 ,$$

hence, by Section 1.2, we have

$$c_1(E|_U) = c_1(L_1|_U) + c_1(L_2|_U) \, .$$

But $H^2(X, \mathbb{Z}) \to H^2(U, \mathbb{Z})$ is an injective map and we obtain the first formula. As we saw above, the holomorphic 2-vector bundle $E \otimes L_1^*$ has a section $s$ with the zero set $Y$. By the example at the end of Section 1.2 we get the formula

$$c_2(E \otimes L_1^*) = [Y].$$

By applying

$$c_2(E \otimes L) = c_2(E) + c_1(E).c_1(L) + c_1^2(L) \, ,$$

where $E$ is a 2-vector bundle and $L$ is a line bundle we get the second formula.

Let $X$ be a compact complex surface. As we saw in Section 3.1 a topological line bundle $E$ admits a holomorphic structure if and only if $c_1(E) \in \mathrm{NS}(X)$. This remains a necessary condition also for $r > 1$ ($r = \mathrm{rank}(E)$), since if $E$ has a holomorphic structure then $\det(E)$ also admits one, hence

$$c_1(E) = c_1(\det(E)) \in \mathrm{NS}(X) \, .$$

Recall that in the surface case the topological classification of vector bundles is known (cf. [Wu]): for every rank $r > 1$ and every pair

$$(c_1, c_2) \in H^2(X, \mathbb{Z}) \times H^4(X, \mathbb{Z})$$

there exists a unique (up to isomorphism) topological complex vector bundle $E$ of rank $r$ with $c_1(E) = c_1$, $c_2(E) = c_2$.

Now we shall present a complete answer to the existence problem for holomorphic (algebraic) vector bundles over algebraic surfaces (see [Sw1]):

**Theorem 4.5 (Schwarzenberger)** *A topological complex vector bundle $E$ over an algebraic surface $X$ admits a holomorphic structure if and only if $c_1(E) \in \mathrm{NS}(X)$.*

*Proof.* We have to construct for given $r > 1$, $c_1 \in \mathrm{NS}(X)$ and $c_2 \in H^4(X, \mathbb{Z}) \cong \mathbb{Z}$, holomorphic vector bundles with $\mathrm{rank}(E) = r$, $c_1(E) = c_1$, $c_2(E) = c_2$. Because we can add a trivial vector bundle of rank $r - 2$, it is enough to consider the case $r = 2$.

We shall apply the Serre method of extensions. Recall that, if $E$ is a holomorphic 2-vector bundle given by an extension

$$0 \to L_1 \to E \to J_Y \otimes L_2 \to 0 \, ,$$

then

$$c_1(E) = c_1(L_1) + c_1(L_2) \, ,$$

$$c_2(E) = c_1(L_1).c_1(L_2) + \ell(Y) \, ,$$

where $\ell(Y)$ is the length of $Y$. Choose now $\mathcal{L} \in \mathrm{Pic}(X)$ such that $c_1(\mathcal{L}) = c_1(E) = c_1 \in \mathrm{NS}(X)$. Let $H$ be an ample line bundle over $X$ ($X$ is algebraic) and choose $L_1 = H^{\otimes n}$, $L_2 = \mathcal{L} \otimes (H^*)^{\otimes n}$. If $n$ is big enough, then

$$H^2(X, L_1 \otimes L_2^*) = H^2(X, \mathcal{L}^* \otimes H^{\otimes(2n)}) = 0$$

and

$$c_1(L_1).c_1(L_2) = n(\mathcal{L}.H) - n^2H^2 \leq c_2 .$$

Then take $Y$ to be a set of $c_2 - (L_1.L_2)$ simple points on $X$. By Theorem 4.3, an extension (4.1) gives a holomorphic 2-vector bundle with given invariants.

*Remark.* As we shall see in the next section the conclusion of the Theorem 4.5 is no longer true when $X$ is a nonalgebraic surface. In this case there are further restrictions on the Chern classes of holomorphic vector bundles.

## 4.2 Filtrable vector bundles

Let $X$ be a complex manifold. See [E-F] for the following

**Definition 4.6** A holomorphic vector bundle $E$ of rank $r$ over the complex manifold $X$ is called *filtrable* if there exists a filtration

$$0 = \mathcal{F}_0 \subset \mathcal{F}_1 \subset \cdots \subset \mathcal{F}_r = E ,$$

where $\mathcal{F}_k$ is a coherent subsheaf of rank $k$.

*Remark.* Of course, every holomorphic (algebraic) vector bundle $E$ over a projective manifold is filtrable. Indeed, let $X$ be a projective manifold and let $H$ be an ample line bundle over $X$. Then $H^0(X, E \otimes H^n) \neq 0$ for $n \gg 0$ (Theorem A). A non-zero section $s \in H^0(X, E \otimes H^n)$ will give an injective homomorphism $\mathcal{O}_X \to E \otimes H^n$ so, an injective homomorphism $H^{-n} \to E$; the image of this homomorphism is a coherent subsheaf of $E$ of rank 1. Take the quotient and repeat the procedure.

As we shall see, on nonalgebraic manifolds there exist holomorphic vector bundles which are not filtrable.

According to the procedure of Definition 1.38, we may assume for a filtrable holomorphic vector bundle that all quotients $E/\mathcal{F}_k$ to be torsion-free. In that case the quotients $\mathcal{F}_k/\mathcal{F}_{k-1}$ are torsion-free of rank 1 and $L_k := (\mathcal{F}_k/\mathcal{F}_{k-1})^{**}$ are line bundles (cf. Proposition 1.33).

**Proposition 4.7** *A holomorphic vector bundle of rank 2 over a connected complex manifold $X$ is filtrable if and only if there exists a line bundle $L$ over $X$ such that $H^0(X, E \otimes L) \neq 0$.*

*Proof.* If $E$ is a holomorphic 2-vector bundle over $X$ which is filtrable, then we have a filtration

$$0 = \mathcal{F}_0 \subset \mathcal{F}_1 \subset E$$

with $\mathcal{F}_1$ a coherent subsheaf of rank 1 and the quotient $E/\mathcal{F}_1$ torsion-free. By Proposition 1.32, $\mathcal{F}_1$ is reflexive, hence it is a line bundle (cf. Proposition 1.33). We get the inclusion $\mathcal{O}_X \hookrightarrow E \otimes \mathcal{F}_1^*$, hence $H^0(X, E \otimes \mathcal{F}_1^*) \neq 0$.

Conversely, let $L$ be a line bundle over $X$ such that $H^0(X, E \otimes L) \neq 0$. By Proposition 4.1 we get an extension

$$0 \to L_1 \to E \to J_Y \otimes L_2 \to 0 \,,$$

with $L_1$, $L_2$ line bundles and $J_Y$ the ideal sheaf of a locally complete intersection $Y$ of codimension 2 in $X$, (or $Y$ is empty set). It follows that there is a filtration

$$0 = \mathcal{F}_0 \subset \mathcal{F}_1 \subseteq E \,,$$

with $\mathcal{F}_1 = \mathrm{Im}(L_1 \to E)$ of rank 1.

Now we can present the following example (see [Ss2]).

**Example 4.8** Let $X$ be a K3-surface with $\mathrm{Pic}(X) = 0$. By Proposition 2.30 we know that $H^0(X, T_X) = 0$, hence by Proposition 4.7 we get that $T_X$ is non-filtrable.

**Definition 4.9** A holomorphic vector bundle $E$ over a compact complex (connected) manifold is *simple* if $\mathrm{End}(E) \cong \mathbb{C}$.

This property is equivalent to the fact that every non-zero endomorphism is invertible (see Proposition 1.35). If $\mathrm{rank}(E) = 2$ and $E$ is not simple, then $E$ is filtrable since, if $\sigma : E \to E$ is a non-zero, non-invertible endomorphism, then $\mathrm{Ker}(\sigma) \subset E$ is a subsheaf of rank 1.

**Lemma 4.10** *Let $E$ be an indecomposable holomorphic 2-vector bundle over a compact (connected) complex manifold $X$ and let $\sigma \in \mathrm{End}(E)$ be a non-invertible endomorphism. Then $\sigma^2 = 0$.*

*Proof.* Consider the eigenvalues $\lambda_1$, $\lambda_2$ of $\sigma$ (since $X$ is compact connected, the eigenvalues of $\sigma$ in all fibres of $E$ are the same). Necessarily $\lambda_1 = \lambda_2$, otherwise the eigenspaces would define a decomposition of $E$. Since $\det(\sigma) = 0$, we have $\lambda_1 = \lambda_2 = 0$, which implies $\sigma^2 = 0$.

**Definition 4.11** An extension

$$0 \to L \to E \to J_Y \otimes M \to 0$$

of a holomorphic 2-vector bundle $E$ is called *maximal*, if for every other extension

$$0 \to L' \to E \to J_Z \otimes M' \to 0 \,,$$

we have a non-zero morphism $L' \to L$.

**Proposition 4.12** *Let $E$ be a non-simple, indecomposable holomorphic 2-vector bundle on a compact connected complex manifold $X$. Then $E$ admits a uniquely determined (up to isomorphism) maximal extension*

$$0 \to L \to E \to J_Y \otimes M \to 0 \,,$$

*characterized by the fact that there is a non-zero morphism $M \to L$.*

*Proof.* (cf. [E-F]) Since $E$ is not simple there exists a non-zero, non-invertible endo-morphism $\sigma : E \to E$. Let $L := \mathrm{Ker}(\sigma)$; since $E/\mathrm{Ker}(\sigma) \cong \mathrm{Im}(\sigma)$ is torsion-free, $L$ is a line bundle by Proposition 1.33. We have $\mathrm{Im}(\sigma) \cong J_Y \otimes M$, where $M$ is a line bundle and $Y \subset X$ is a subspace of codimension 2 (or empty). By Lemma 4.10 we have $\sigma^2 = 0$, hence $\mathrm{Im}(\sigma) \subset \mathrm{Ker}(\sigma)$. It follows that there exists a monomorphism $J_Y \otimes M \to L$, which extends to a monomorphism $M \to L$. Thus we get an extension

$$0 \to L \xrightarrow{\alpha} E \xrightarrow{\beta} J_Y \otimes M \to 0$$

with a non-zero morphism $M \to L$.

Let $f : L' \to E$ be any non-zero morphism. If $\beta \circ f : L' \to J_Y \otimes M$ is non-zero, then the morphism $L' \to M \to L$ is non-zero. If $\beta \circ f = 0$, we have:

$$L' \cong \mathrm{Im}(f) \subset \mathrm{Im}(\alpha) \cong L \ ,$$

hence the morphism $L' \to L$ is again non-zero. This shows that the above extension is maximal.

Let

$$0 \to L' \xrightarrow{f} E \to J_{Y'} \otimes M' \to 0$$

be a second maximal extension. Then we have non-zero morphisms $L' \to L$ and $L \to L'$; it follows $L' \cong L$. If $\beta \circ f : L' \to J_Y \otimes M$ is non-zero, the composite map

$$L' \xrightarrow{\beta \circ f} J_Y \otimes M \to M \to L$$

is non-zero, hence it is an isomorphism. This implies $Y = \emptyset$ and $\beta \circ f : L' \to M$ is an isomorphism. Then $E \cong L \oplus L'$, contradiction. It follows $\beta \circ f = 0$ and we get a factorization

$$
\begin{array}{ccc}
 & L' & \\
g \swarrow & & \searrow f \\
L & \xrightarrow{\alpha} & E
\end{array}
$$

Since $L' \cong L$, $g$ is an isomorphism. This implies that the two extensions are isomorphic.

We shall see now that, in the nonalgebraic case, the condition $c_1(E) \in \mathrm{NS}(X)$ from Theorem 4.5 is no longer sufficient for the existence of holomorphic structures in a topologically complex 2-vector bundle $E$ (see [E-F, B-F1]):

**Proposition 4.13** *Let $X$ be a nonalgebraic surface and let $a \in \mathrm{NS}(X)$ be fixed. Then, for every holomorphic 2-vector bundle $E$ over $X$ with $c_1(E) = a$, we have*

$$c_2(E) \geq \min \left\{ \frac{1}{4} a^2, \ 2\chi(\mathcal{O}_X) + \frac{1}{2}(c_1(X).a + a^2) \right\} \ .$$

*Proof.* By Riemann-Roch Theorem we get

$$\chi(E) = 2\chi(\mathcal{O}_X) + \frac{1}{2}(c_1(X).a + a^2) - c_2(E) \ .$$

If $E$ is non-filtrable then, by Proposition 4.7, it follows

$$H^0(X, E) = 0, \ H^2(X, E) \cong H^0(X, E^* \otimes \mathcal{K}_X)^* = 0$$

(because $E \cong E^* \otimes \det(E)$ implies $H^0(X, E^* \otimes \mathcal{K}_X) \cong H^0(X, E \otimes \mathcal{K}_X \otimes \det(E^*)))$, hence $\chi(E) = -h^1(X, E) \leq 0$. Therefore

$$c_2(E) \geq 2\chi(\mathcal{O}_X) + \frac{1}{2}(c_1(X).a + a^2) \ .$$

If $E$ is filtrable then, from the extension (4.1), by Proposition 4.4, we have

$$a = c_1(L_1) + c_1(L_2), \ c_2(E) = c_1(L_1).c_1(L_2) + \ell(Y) \ .$$

The discriminant of $E$ is given by

$$\Delta(E) = \frac{1}{2}\left(c_2(E) - \frac{1}{4}c_1^2(E)\right) = \frac{1}{8}(\ell(Y) - c_1^2(L_1 \otimes L_2^*)) \ .$$

Since $X$ is nonalgebraic $c_1^2(L_1 \otimes L_2^*) \leq 0$, hence $\Delta(E) \geq 0$, i.e.

$$c_2(E) \geq \frac{1}{4}a^2 \ .$$

**Corollary 4.14** *Let $X$ be a nonalgebraic surface. Then there exist topological complex 2-vector bundles $E$ over $X$ with holomorphic structure on $\det(E)$, but which do not admit any holomorphic structure.*

For more general results we need some preliminaries. Recall that we defined in Section 1.2 the Chern classes and the rank for any analytic coherent sheaf $\mathcal{F}$ over $X$.

**Lemma 4.15** *The Chern class $c_1(\mathcal{F})$ belongs to the Neron-Severi group $NS(X)$ for every analytic coherent sheaf $\mathcal{F}$ over a complex surface $X$.*

*Proof.* If $\mathcal{F}$ is locally free, then $c_1(\mathcal{F}) = c_1(\det(\mathcal{F})) \in NS(X)$. A result of Schuster (see [Ss2]) says that any analytic coherent sheaf $\mathcal{F}$ over a complex surface has a resolution

$$0 \to E_2 \to E_1 \to E_0 \to \mathcal{F} \to 0 \ ,$$

with $E_i$ locally free sheaves. Then

$$c_1(\mathcal{F}) = \sum_i (-1)^i c_1(E_i) \in NS(X) \ .$$

**Definition 4.16** Let $\mathcal{F}$ be an analytic coherent sheaf over a surface $X$ of rank $r > 0$, with Chern classes $c_1(\mathcal{F})$ and $c_2(\mathcal{F})$ The *discriminant* $\Delta(\mathcal{F})$ is defined by

$$\Delta(\mathcal{F}) := \frac{1}{r}\left(c_2(\mathcal{F}) - \frac{r-1}{2r}c_1^2(\mathcal{F})\right) .$$

Remark that if $L$ is a holomorphic line bundle over $X$, then

$$\Delta(\mathcal{F} \otimes L) = \Delta(\mathcal{F}) ,$$

since we have

$$c_1(\mathcal{F} \otimes L) = c_1(\mathcal{F}) + rc_1(L)$$

$$c_2(\mathcal{F} \otimes L) = c_2(\mathcal{F}) + (r-1)c_1(\mathcal{F}).c_1(L) + \binom{r}{2}c_1^2(L) .$$

Bănică and Le Potier (see [B-L]) obtained the following more general result:

**Theorem 4.17** *Let $X$ be a nonalgebraic surface and $\mathcal{F}$ a torsion-free coherent sheaf over $X$ of rank $r$, with Chern classes $c_1(\mathcal{F})$ and $c_2(\mathcal{F})$. Then $\Delta(\mathcal{F}) \geq 0$.*

*Proof.* (V. Vuletescu) This elementary proof is from [Vu2]. Firstly, we shall prove the following inequality: for every torsion-free coherent sheaf $\mathcal{F}$ of rank $r$ over $X$ we have

$$\chi(\mathcal{F}) \leq r\chi(\mathcal{O}_X). \tag{4.2}$$

*Step 1.* We shall prove that if $L$ is a line bundle over a minimal surface $X$, then

$$c_1^2(L) \leq c_1(L).c_1(\mathcal{K}_X).$$

From Kodaira classification of nonalgebraic surfaces we know that always $\chi(\mathcal{O}_X) \geq 0$. Assume $L$ would be some line bundle for which

$$c_1^2(L) > c_1(L).c_1(\mathcal{K}_X) .$$

By Riemann-Roch formula we get $\chi(L) > 0$, hence $L$ or $L^* \otimes \mathcal{K}_X$ is effective. We shall consider only the first case, the second case being similar. One has $L \cong \mathcal{O}_X(\sum m_i C_i)$, with $m_i > 0$ and $C_i$ irreducible curves on $X$. Since $c_1^2(L) \leq 0$ (by Corollary 2.9) we get $c_1^2(L) < 0$ (otherwise, by Corollary 2.10, we get a contradiction with the assumption we made). But in this case, at least one of the curves $C_i$ would satisfy

$$C_i^2 < 0 \text{ and } (C_i.\mathcal{K}_X) < 0 ,$$

hence it would be a (-1)-curve (cf. [B-P-V], p. 72), contradiction with the minimality assumption. The step 1 is proved.

*Step 2.* We shall prove the same inequality for any nonalgebraic surface $X$. Apply induction on the length $n$ of a chain of blow-downs which produce a minimal $X'$, bimeromorphic to $X$. For $n = 0$, this has been proved at step 1, so assume $X$ has a chain of blow-downs of length $n > 0$. Denote by $X_1$ the surface obtained by a single blow-down, say $g : X \to X_1$. As $X_1$ has a chain of length $n - 1$, the induction hypothesis applies to it. Let $L \in \text{Pic}(X)$ be arbitrary; by Theorem 2.20 there exist and there are unique a line bundle $L_1 \in \text{Pic}(X_1)$ and $m \in \mathbb{Z}$ such that

$$L \cong g^*(L_1) \otimes \mathcal{O}_X(mE) ,$$

where $E$ is the exceptional divisor of $g$. As

$$c_1^2(L) - c_1(L).c_1(\mathcal{K}_X) = c_1^2(L_1) - c_1(L_1).c_1(\mathcal{K}_{X_1}) - m(m-1) ,$$

and since

$$c_1^2(L_1) - c_1(L_1).c_1(\mathcal{K}_{X_1}) \leq 0$$

by the induction hypothesis, the proof of step 2 is over.

Thus, we proved

$$\chi(L) \leq \chi(\mathcal{O}_X)$$

for every line bundle over a nonalgebraic surface $X$ (look at Riemann-Roch formula for $L$).

*Step 3.* We show that the inequality (4.2) holds for torsion-free sheaves of rank 1. To see this, recall (Section 1.4) that for every torsion-free sheaf $L$ one has the exact sequence

$$0 \to L \to L^{**} \to Q \to 0 ,$$

where the sheaf $Q$ has the support of codimension at least two. It follows $\chi(Q) \geq 0$ ($\chi(Q)$ = length of the support of $Q$), hence

$$\chi(L) \leq \chi(L^{**}) \leq \chi(\mathcal{O}_X)$$

($L^{**}$ being a line bundle by Proposition 1.33).

*Step 4.* This is the final step. We proceed by induction on the rank $r$ of the torsion-free sheaf $\mathcal{F}$. For $r = 1$, the proof has been already done, so assume we proved the statement for all torsion-free sheaves of rank $\leq r - 1$. Let firstly $\mathcal{F} = E$ be a holomorphic vector bundle of rank $r$. If $E$ has no coherent subsheaf, then $H^0(X, E) = 0$, since non-zero global sections in $E$ would generate rank 1 coherent subsheaves. Similarly,

$$H^2(X, E) \cong H^0(X, E^* \otimes \mathcal{K}_X) = 0 ,$$

otherwise non-zero global sections in $E^* \otimes \mathcal{K}_X$ would produce rank 1 coherent quotients of $E$. Then

$$\chi(E) \leq 0 \leq r\chi(\mathcal{O}_X) .$$

If $E$ has some coherent subsheaf we get the following exact sequence

$$0 \to L \to E \to Q \to 0 ,$$

with $L$ and $Q$ torsion-free sheaves of ranks at most $r - 1$. By the induction hypothesis, we are done. Finally, let $\mathcal{F}$ be a torsion-free sheaf of rank $r$. Using again the exact sequence

$$0 \to \mathcal{F} \to \mathcal{F}^{**} \to Q \to 0 ,$$

and the arguments from step 3, we finish the proof of the inequality

$$\chi(\mathcal{F}) \leq r\chi(\mathcal{O}_X) .$$

Let $\mathcal{F}$ be a torsion-free coherent sheaf of rank $r > 1$ over the nonalgebraic surface $X$. The bidual sheaf $\mathcal{F}^{**}$ is locally free (see Section 1.4) and we have $\Delta(\mathcal{F}^{**}) \leq \Delta(\mathcal{F})$.

Thus, we reduced the problem to the case of a holomorphic vector bundle $E = \mathcal{F}^{**}$. By the Riemann-Roch Theorem we have

$$\chi(E) = \frac{1}{2}c_1(E).(c_1(E) - c_1(\mathcal{K}_X)) - c_2(E) + r\chi(\mathcal{O}_X) .$$

If $c_1(E) = 0$, by the inequality proved above, we get $c_2(E) \geq 0$. Now apply this fact to the vector bundle $sl(E)$ of trace-free endomorphisms of $E$, taking into accout that $c_1(sl(E)) = 0$ and

$$c_2(sl(E)) = 2rc_2(E) - (r-1)c_1^2(E) .$$

We obtain $\Delta(E) \geq 0$.

*Remark.* In fact, the inequality $\chi(\mathcal{F}) \leq r\chi(\mathcal{O}_X)$ is equivalent to the inequality $\Delta(\mathcal{F}) \geq 0$, for any torsion-free coherent sheaf $\mathcal{F}$ over $X$.

**Corollary 4.18** *Let $X$ be a nonalgebraic surface. Then there exist topological vector bundles $E$ of any rank $r \geq 2$ over $X$ with holomorphic structure on $\det(E)$, but which do not admit any holomorphic structure.*

We shall present a complete answer, obtained by Bănică-Le Potier [B-L], to the existence problem for filtrable holomorphic vector bundles over nonalgebraic surfaces; for the case of rank 2-vector bundles see [LP2] for K3-surfaces, and [B-F1] for 2-tori. We shall follow closely [B-L].

Let $\mathcal{F}$ be an analytic coherent sheaf of rank $r > 0$, with Chern classes $c_1(\mathcal{F})$ and $c_2(\mathcal{F})$ over a nonalgebraic surface $X$.

**Definition 4.19** The *slope* $\mu(\mathcal{F})$ of the sheaf $\mathcal{F}$ is defined by

$$\mu(\mathcal{F}) := \frac{c_1(\mathcal{F})}{r} \in \text{NS}(X) \otimes \mathbb{Q} .$$

We shall denote by $K$ the Chern class of the canonical line bundle $\mathcal{K}_X$. The Riemann-Roch Theorem can be written in the following form

$$\chi(\mathcal{F}) = r(P(\mu(\mathcal{F})) - \Delta(\mathcal{F})) ,$$

where

$$P : \text{NS}(X) \otimes \mathbb{Q} \longrightarrow \mathbb{Q}$$

is the polynomial

$$P(\mu) = \chi(\mathcal{O}_X) - \frac{1}{2}\mu.K + \frac{1}{2}\mu^2 = \chi(\mathcal{O}_X) - \frac{1}{8}K^2 + \frac{1}{2}(\mu - K/2)^2 .$$

Since, by the Corollary 2.9 (Kodaira), we have that the intersection form is negative semi-definite on $\text{NS}(X) \otimes \mathbb{Q}$ for a nonalgebraic surface $X$, it follows that the function

$$\mu \mapsto P(\mu) : \text{NS}(X) \otimes \mathbb{Q} \longrightarrow \mathbb{Q}$$

is concave.

**Definition 4.20** For a nonalgebraic surface $X$, $a \in NS(X)$ and $r$ a positive integer we can define the following rational positive number

$$m(r,a) := -\frac{1}{2r}\max \left\{ \sum_{i=1}^{r} \left( \frac{a}{r} - \mu_i \right)^2 \mid \mu_1, \cdots, \mu_r \in NS(X) \text{ with } \sum_{i=1}^{r} \mu_i = a \right\} .$$

*Remark.* When $X$ is a 2-torus and $r = 2$ an explicit description of the invariants $m(2,a)$ is given in [B-F1].

We have (see [B-L]):

**Theorem 4.21 (Bǎnicǎ-Le Potier)** *A rank $r \geq 2$ topological complex vector bundle $E$ over a nonalgebraic surface $X$ admits a filtrable holomorphic structure if and only if*

$$c_1(E) \in NS(X) \text{ and } \Delta(E) \geq m(r, c_1(E)) ,$$

*except when $X$ is a K3-surface with $a(X) = 0$, $c_1(E) \in rNS(X)$ and $\Delta(E) = 1/r$. In this excepted case $E$ admits no holomorphic structures.*

*Proof.* (cf. [B-L]) We shall prove firstly that there exists a filtrable coherent sheaf of rank $r \geq 2$, with Chern classes $c_1 \in NS(X)$ and $c_2 \in \mathbb{Z}$ over $X$ if and only if $\Delta \geq m(r, c_1)$. Let $\mathcal{F}$ be a filtrable coherent sheaf of rank $r \geq 2$, with Chern classes $c_1$ and $c_2$. Then, there exists a filtration

$$0 = \mathcal{F}_0 \subset \mathcal{F}_1 \subset \cdots \subset \mathcal{F}_r = \mathcal{F}$$

such that

$$gr_i := \mathcal{F}_i/\mathcal{F}_{i-1} , \ i = 1, \cdots, r$$

are torsion-free sheaves of rank one (see Definition 1.38). It follows that $gr_i^{**}$ are line bundles (cf. Proposition 1.33). Let $\mu$, $\Delta$ be the slope and the discriminant of $\mathcal{F}$ and let $\mu_i$, $\Delta_i$ be the slope and the discriminant of $gr_i$. Since we have

$$\chi(\mathcal{F}) = \sum_i \chi(gr_i) ,$$

it follows by the Riemann-Roch formula that

$$\Delta = P(\mu) - \frac{\chi(\mathcal{F})}{r} = P(\mu) - \frac{1}{r}\sum_i P(\mu_i) + \frac{1}{r}\sum_i \Delta_i$$

or

$$\Delta = \frac{1}{2}\left( \mu^2 - \frac{1}{r}\sum_i \mu_i^2 \right) + \frac{1}{r}\sum_i \Delta_i . \tag{4.3}$$

Since $gr_i$ is of rank one, $\mu_i \in NS(X)$ and $\sum_i \mu_i = c_1$. We have the exact sequences

$$0 \to gr_i \to gr_i^{**} \to Q_i \to 0 ,$$

with $Q_i$ analytic coherent sheaves with 0-dimensional supports (see Corollary 1.25). It follows that

$$\chi(gr_i^{**}) = \chi(gr_i) + h^0(X, Q_i) \ .$$

By Riemann-Roch formula for $gr_i$ and $gr_i^{**}$ we get $\Delta_i = h^0(X, Q_i) \geq 0$. But

$$\frac{1}{2}\left(\mu^2 - \frac{1}{r}\sum_i \mu_i^2\right) = \frac{1}{2}\left(\left(\frac{c_1}{r}\right)^2 - \frac{1}{r}\sum_i \mu_i^2\right) =$$

$$= -\frac{1}{2r}\sum_i \left(\frac{c_1}{r} - \mu_i\right)^2 \geq m(r, c_1) \ ,$$

hence

$$\Delta(\mathcal{F}) \geq m(r, c_1) \ .$$

Conversely, let $r$ be an integer $\geq 2$, $c_1 \in \text{NS}(X)$, $c_2 \in \mathbb{Z}$ and suppose that

$$\Delta \geq m(r, c_1) \ .$$

Choose elements $\mu_1, \cdots, \mu_r \in \text{NS}(X)$ such that

$$\sum_i \mu_i = c_1 \ ,$$

$$\frac{1}{2}\left(\left(\frac{c_1}{r}\right)^2 - \frac{1}{r}\sum_i \mu_i^2\right) = m(r, c_1) \ .$$

Then, with the notation $\mu = c_1/r$, we have

$$r(\Delta - m(r, c_1)) = r(\Delta - P(\mu)) + \sum_i P(\mu_i) \ .$$

One verifies by direct computation that

$$r(\Delta - P(\mu)) \in \mathbb{Z} \text{ and } P(\mu_i) \in \mathbb{Z} \ .$$

It follows that

$$\ell = r(\Delta - m(r, c_1)) \in \mathbb{Z}, \quad \ell \geq 0 \ .$$

Now, choose for $i = 1, \cdots, r$ any holomorphic line bundle $L_i$ with $c_1(L_i) = \mu_i$, and take $Y \subset X$ an analytic subspace with finite support, of length $\ell$. If $J_Y$ is the ideal sheaf of $Y$, then the analytic coherent sheaf

$$\mathcal{F} = L_1 \oplus \cdots \oplus L_{r-1} \oplus (J_Y \otimes L_r)$$

has rank $r$ and the first Chern class $c_1(\mathcal{F}) = c_1$. By the formula (4.3) we get that the discriminant of $\mathcal{F}$ is $\Delta$, hence $c_2(\mathcal{F}) = c_2$.

Now, we consider the case of filtrable holomorphic vector bundles. If $\Delta = m(r, c_1)$ then $\ell = 0$ and $Y$ is the empty set. It follows that

$$\mathcal{F} = L_1 \oplus \cdots \oplus L_{r-1} \oplus L_r$$

is a filtrable holomorphic structure in the topological complex vector bundle $E$.

If $\Delta > m(r, c_1)$ we denote

$$\mathcal{F}_1 = L_{r-1} \oplus (J_Y \otimes L_r) \ ;$$

$\mathcal{F}_1$ is an analytic coherent sheaf of rank 2 but not a vector bundle. From the exact sequence

$$0 \to L_{r-1} \to \mathcal{F}_1 \to J_Y \otimes L_r \to 0$$

we get

$$c_1(\mathcal{F}_1) = \mu_{r-1} + \mu_r, \quad c_2(\mathcal{F}_1) = \ell + \mu_{r-1} \cdot \mu_r .$$

Then

$$\Delta(\mathcal{F}_1) = \frac{1}{2}\left(\ell - \frac{1}{2}\left(\frac{c_1(\mathcal{F}_1)}{2} - \mu_{r-1}\right)^2\right) .$$

It follows that it suffices to prove that there exists a filtrable holomorphic 2-vector bundle $E'$ with the same Chern classes as $\mathcal{F}_1$, hence we reduced the problem to the case $r = 2$ ($\mathcal{F}' = L_1 \oplus \cdots \oplus L_{r-2} \oplus E'$ will be the solution for rank $r$ case).

Suppose $r = 2$ and choose $\mu_1, \mu_2 \in \mathrm{NS}(X)$ such that

$$\mu_1 + \mu_2 = c_1 , \quad -\frac{1}{2}\left(\frac{c_1}{2} - \mu_1\right)^2 = m(2, c_1) . \tag{4.4}$$

If there exists an irreducible curve $C$ on $X$ we can suppose moreover that $\mu_1.C \geq \mu_2.C$ (changing, if necessary, $\mu_1$ with $\mu_2$). Now, we take as above, holomorphic line bundles $L_1, L_2$ with $c_1(L_1) = \mu_1$, $c_1(L_2) = \mu_2$ and an analytic subspace $Y \subset X$ with finite support of length $\ell = 2(\Delta - m(2, c_1)) > 0$ and we apply Serre method (Theorem 4.3). Thus, we try to find extensions

$$0 \to L_1 \to E' \to J_Y \otimes L_2 \to 0 ,$$

with $E'$ a holomorphic 2-vector bundle. Let $L$ be $L_1 \otimes L_2^*$; then these extensions are classified by $\mathrm{Ext}^1(J_Y, L)$ and we have the exact sequence

$$\mathrm{Ext}^1(J_Y, L) \xrightarrow{\alpha} \mathrm{Ext}^2(\mathcal{O}_Y, L) \xrightarrow{\beta} H^2(X, L) , \tag{4.5}$$

where the last map is the dual of the canonical restriction map

$$H^0(X, L^* \otimes K_X) \longrightarrow H^0(X, L^* \otimes K_X \otimes \mathcal{O}_Y) \tag{4.6}$$

(see the considerations in the proof of Theorem 4.3).

(a) *The case $X$ non-minimal.* There exists on $X$ a smooth rational curve $C$ such that $C^2 = -1$ (a $(-1)$-curve) and we choose $\mu_1, \mu_2 \in \mathrm{NS}(X)$ with the property $\mu_1.C \geq \mu_2.C$. From the genus formula

$$2(g - 1) = C.K + C^2 ,$$

it follows that the line bundle $K_X|_C$ has degree $-1$ and, since $(\mu_2 - \mu_1).C \leq 0$, we get that the line bundle $(L^* \otimes K_X)|_C$ has negative degree and the sections of the line bundle $L^* \otimes K_X$ vanish on the curve $C$. If we choose $Y \subset C$, locally complete intersection of length $\ell$, then the restriction map (4.6) is zero, hence the map $\beta$ is zero. From the exact sequence (4.5) we get the surjective map

$$\mathrm{Ext}^1(J_Y, L) \xrightarrow{\alpha} H^0(Y, \mathcal{O}_Y) \to 0 ,$$

and any element $\xi \in \mathrm{Ext}^1(J_Y, L)$ with image $1 \in H^0(Y, \mathcal{O}_Y)$ will give a filtrable holomorphic 2-vector bundle $E'$.

(b) *The case $a(X) = 1$.* In this case $X$ is an elliptic surface and let $f : X \to S$ be a connected surjective morphism onto a curve $S$ (see Theorem 2.13). The image of the natural morphism

$$f^* : H^2(S, \mathbb{Q}) \longrightarrow \mathrm{NS}(X) \otimes \mathbb{Q}$$

is totally isotropic by Corollary 2.10 and Theorem 2.38. It follows that the conditions (4.4) are still satisfied if we replace $\mu_1$ by $\mu_1' = \mu_1 + f^*(\alpha)$ and $\mu_2$ by $\mu_2' = \mu_2 - f^*(\alpha)$ with $\alpha \in H^2(S, \mathbb{Z})$. Let $A$ be a line bundle over $S$ with Chern class $c_1(A) = \alpha$ and take

$$L_1' = L_1 \otimes f^*(A) , \quad L_2' = L_2 \otimes f^*(A^*) ,$$
$$L' = L_2'^* \otimes L_1' = L \otimes f^*(A^{\otimes 2}) .$$

We have

$$h^0(X, L'^* \otimes K_X) = h^0(X, L^* \otimes K_X \otimes f^*(A^{*\otimes 2})) = h^0(S, f_*(L^* \otimes K_X) \otimes A^{*\otimes 2}) .$$

The sheaf $f_*(L^* \otimes K_X)$ has no torsion, hence it is locally free over $S$. If we choose $A$ sufficiently ample then

$$h^0(S, f_*(L^* \otimes K_X) \otimes A^{*\otimes 2}) = 0 .$$

Thus, if we choose $Y$ arbitrarily, a locally complete intersection of length $\ell$, then we get an extension

$$0 \to L_1' \to E' \to J_Y \otimes L_2' \to 0 ,$$

where $E'$ is a (filtrable) holomorphic 2-vector bundle with Chern classes $c_1$ and $c_2$.

(c) *The case $a(X) = 0$.* By Proposition 2.15 we know that $h^0(X, L^* \otimes K_X) \leq 1$. If $h^0(X, L^* \otimes K_X) = 0$ then we can choose $Y$ an arbitrarily locally complete intersection of length $\ell$ and there are extensions as above with $E'$ a (filtrable) holomorphic 2-vector bundle. If $h^0(X, L^* \otimes K_X) = 1$ and if $L \neq K_X$, then the holomorphic sections of the line bundle $L^* \otimes K_X$ vanish on a curve $C$. One chooses $Y \subset C$, locally complete intersection of length $\ell$ and we obtain as above 2-vector bundles $E'$. If $L \cong K_X$ and if $q = h^1(X, \mathcal{O}_X) \neq 0$, then the group of topologically trivial line bundles

$$\mathrm{Pic}_0(X) \cong H^1(X, \mathcal{O}_X)/H^1(X, \mathbb{Z})$$

is not zero. Let $A$ be a non-trivial element of $\mathrm{Pic}_0(X)$. Then we change $L_1$ by $L_1' = L_1 \otimes A$ (which does not change the Chern class $c_1(L_1') = \mu_1$) and we have $L_2^* \otimes L_1' = L \otimes A \neq K_X$. Now we can use the previous construction.

It remains the case $a(X) = 0$ and $q(X) = 0$. If $X$ is not minimal we use case (a). If $X$ is minimal then, by the classification Theorem 2.25, we get that $X$ is a K3-surface.

(d) *The case of K3-surfaces with $a(X) = 0$.* From the Theorem 2.37 we know that the quadratic intersection form is negative-definite on the group $\mathrm{NS}(X) \cong \mathrm{Pic}(X)$, which has no torsion.

Let $c_1 \in \mathrm{NS}(X)$ with $c_1 = 0 \bmod r$. Let us suppose that there exists over $X$ a filtrable holomorphic vector bundle $E$ of rank $r$, with Chern classes $c_1$ and $c_2$. We shall prove that $\Delta \neq 1/r$. By tensoring $E$ to a suitable line bundle one can suppose $c_1 = 0$. Let

$$0 = \mathcal{F}_0 \subset \mathcal{F}_1 \subset \cdots \subset \mathcal{F}_r = E$$

be a filtration of $E$ such that $gr_i = \mathcal{F}_i/\mathcal{F}_{i-1}$, $i = 1, \cdots, r$ are torsion-free of rank one. From the formula (4.3) we get

$$c_2 = r\Delta = -\frac{1}{2}\sum_i \mu_i^2 + \sum_i \Delta_i \; ,$$

$$\sum_i \mu_i = 0 \; .$$

If not all $\mu_i$ are zero, then at least two of them are non-zero. Since the intersection form on $\mathrm{NS}(X)$ is even and negative-definite, we get $\mu_i^2 \leq -2$ for $\mu_i \neq 0$. It follows that $\sum_i \mu_i^2 \leq -4$, hence $c_2 = r\Delta \geq 2$. If all $\mu_i$ are zero, then we get $c_2 = r\Delta = \sum_i \Delta_i$. If $c_2 = r\Delta = 1$, then only one sheaf $gr_i$ is not locally free. It follows that $gr_i = \mathcal{O}_X$ for $i \neq i_0$ and $gr_{i_0}$ is the ideal sheaf $J_Y$ of a simple point $Y \subset X$. Then

$$\mathrm{Ext}^1(gr_i, gr_j) = 0 \quad \text{for } i \neq j \; .$$

From the exact sequences

$$0 \to \mathcal{F}_{i-1} \to \mathcal{F}_i \to gr_i \to 0$$

we deduce that the filtration $(\mathcal{F}_i)$ is splitable, i.e.

$$E \cong \oplus_i gr_i \; .$$

But then $E$ is not locally free, contradiction.

Conversely, suppose firstly that $\Delta \geq m(r, c_1)$ and $c_1 \neq 0 \bmod r$. Then the classes $\mu_i$ (chosen as above) are not all equal. We reduce the problem to the case $r = 2$ as above. In this case we have $\mu_1 \neq \mu_2$. For $L_1, L_2$ line bundles with $c_1(L_1) = \mu_1, c_1(L_2) = \mu_2$ we get that the line bundle $L = L_2^* \otimes L_1$ is not trivial. But the canonical bundle $\mathcal{K}_X \cong \mathcal{O}_X$ is trivial, so $L \neq \mathcal{K}_X$ and we can apply the argument used in the case (c).

If $\Delta \geq m(r, c_1)$ and $c_1 = 0 \bmod r$ we choose $\mu_i = c_1/r$ for all $i = 1, \cdots, r$. If we suppose $\Delta \neq 1/r$, then it suffices to prove that for a holomorphic line bundle $L$ with Chern class $c_1(L) = c_1/r$ and $Y \subset X$ a locally complete intersection of length $\ell \neq 1$, there exists a filtrable holomorphic vector bundle with the same Chern classes as the sheaf

$$\underbrace{L \oplus \cdots \oplus L}_{r-1} \oplus (J_Y \otimes L) \; .$$

The case $\ell = 0$ is trivial. For $\ell > 1$ we reduce again the problem to the case $r = 2$ and we consider extensions of the form

$$0 \to \mathcal{O}_X \to E' \to J_Y \to 0$$

(since we can tensor with $L^*$). We can find $E'$ locally free if we can find an element $\theta' \in \mathrm{Ext}^2(\mathcal{O}_Y, \mathcal{O}_X) \cong \mathbb{C}^\ell$, which generates the sheaf $\mathcal{E}xt^2(\mathcal{O}_Y, \mathcal{O}_X)$ and whose image in $H^2(X, \mathcal{O}_X)$ is zero. We choose $Y$ to be a union of $\ell$ distinct simple points; then the kernel of the canonical map

$$\mathrm{Ext}^2(\mathcal{O}_Y, \mathcal{O}_X) \longrightarrow H^2(X, \mathcal{O}_X)$$

is a hyperplane, which is not contained in none of the hyperplanes $H_y = \mathrm{Ker}\ \rho_y$, where

$$\rho_y : \operatorname{Ext}^2(\mathcal{O}_Y, \mathcal{O}_X) \longrightarrow \mathcal{E}xt^2(\mathcal{O}_Y, \mathcal{O}_X)_y \cong \mathbb{C} \ ,$$

are canonical morphisms for any $y \in Y$. This fact follows by duality and the proof is over.

**Corollary 4.22** *Let $X$ be a nonalgebraic surface, $r \geq 2$ an integer, $c_1 \in NS(X)$ and $c_2 \in \mathbb{Z}$. Suppose $m(r, c_1) = 0$. Then there exists a holomorphic vector bundle over $X$ of rank $r$, with Chern classes $c_1$ and $c_2$ if and only if $\Delta \geq 0$, except when $X$ is a K3-surface with $a(X) = 0$ and $\Delta = 1/r$.*

*Remarks* (1) For the numbers $m(r, c_1)$ we have the inequalities:

$$-\frac{1}{2} \sup_{\mu \in NS(X)} \left( \frac{c_1}{r} - \mu \right)^2 \leq m(r, c_1) \leq -\frac{r-1}{2} \sup_{\mu \in NS(X)} \left( \frac{c_1}{r} - \mu \right)^2 \ .$$

The first inequality is trivial and the second is obtained by choosing $\mu \in NS(X)$ with $(c_1/r - \mu)^2$ maximal and $\mu_1 = \cdots = \mu_{r-1} = \mu$, $\mu_r = c_1 - (r-1)\mu$. In the case $r = 2$ we get

$$m(2, c_1) = -\frac{1}{2} \sup_{\mu \in NS(X)} \left( \frac{c_1}{2} - \mu \right)^2 \ .$$

(2) The condition $m(r, c_1) = 0$ is equivalent with the existence of an element $\mu \in NS(X)$ such that $(c_1/r - \mu)^2 = 0$. In particular, this condition is satisfied for $c_1 = 0 \bmod r$ or $c_1^2 = 0$ (by Corollary 2.10). More particularly, it is satisfied if $c_1$ is a torsion class, for example if $b_2(X) = 0$. This is the case for Hopf surfaces, Inoue-Bombieri surfaces and secondary Kodaira surfaces. We have the following interesting result (see [B-L]), which says that the existence of holomorphic structures in the case of nonalgebraic surfaces strongly depends on the rank:

**Corollary 4.23** *Let $X$ be a nonalgebraic surface and let $r \geq 2$ be an integer. Suppose that the quadratic intersection form $NS(X) \to \mathbb{Z}$ is not zero. Then there exists over $X$ a topological vector bundle $E$ of rank $r$ such that $E \oplus \mathbb{1}$ has a holomorphic structure, but $E$ has none.*

*Proof.* Let $e \in NS(X)$ with $e^2 < 0$. Consider the topological vector bundle $E$ of rank $r$, with Chern classes $c_1 = (r+1)e$ and $c_2$. Then $m(r+1, c_1) = 0$ ($c_1 = 0 \bmod (r+1)$). We choose $c_2 \in \mathbb{Z}$ such that

$$c_2 - \frac{r-1}{2r}c_1^2 < 0 \ ; \ c_2 - \frac{r}{2(r+1)}c_1^2 \geq 0 \ \text{and} \neq 1 \ ,$$

i.e.

$$d \leq c_2 < (1 - \frac{1}{r^2})d \ , \ c_2 \neq d+1 \ ,$$

where $d = \frac{1}{2}r(r+1)e^2$. It follows by the Theorem 4.17 that $E$ has no holomorphic structures and by the Theorem 4.21 that $E \oplus \mathbb{1}$ has filtrable holomorphic structures (here $\mathbb{1}$ denotes as usual the topologically trivial line bundle).

## 4.3  Non-filtrable and irreducible vector bundles

As we have seen in the Example 4.8 there exist non-filtrable holomorphic vector bundles over nonalgebraic surfaces. This is an essential difference between the algebraic and nonalgebraic case. Moreover, we can give the following

**Definition 4.24** A holomorphic vector bundle $E$ over a complex (connected) manifold $X$ is called *reducible* if it admits a coherent analytic subsheaf $\mathcal{F}$ such that

$$0 < \text{rank}(\mathcal{F}) < \text{rank}(E) \,.$$

and *irreducible* otherwise.

Of course, if $E$ is of rank 2, then $E$ is reducible if and only if $E$ is filtrable. By Proposition 4.7, we get that a holomorphic 2-vector bundle is irreducible if and only if $H^0(X, E \otimes L) = 0$ for every $L \in \text{Pic}(X)$.

**Proposition 4.25** *A holomorphic vector bundle $E$ of rank 3 over a complex (connected) manifold $X$ is irreducible if and only if $H^0(X, E \otimes L) = 0$ and $H^0(X, E^* \otimes L) = 0$ for every $L \in \text{Pic}(X)$.*

*Proof.* Let $E$ be irreducible and suppose that $H^0(X, E \otimes L) \neq 0$ for some $L \in \text{Pic}(X)$. A non-zero section of $H^0(X, E \otimes L)$ will give an injective homomorphism

$$0 \to L^* \to E$$

and its image will be a coherent subsheaf of $E$ of rank 1, contradiction. Now, suppose that $H^0(X, E^* \otimes L) \neq 0$ for some $L \in \text{Pic}(X)$. Then the injective homomorphism $L^* \to E^*$ given by a non-zero section gives a non-zero morphism $E \to L$. It follows that the rank of the image of this last morphism is 1 and the kernel of the morphism will be a coherent subsheaf of $E$ of rank 2, contradiction.

Conversely, suppose that

$$H^0(X, E \otimes L) = H^0(X, E^* \otimes L) = 0$$

for every $L \in \text{Pic}(X)$. Assume that $E$ is reducible and let $\mathcal{F} \subset E$ be a coherent subsheaf with

$$0 < \text{rank}(\mathcal{F}) < \text{rank}(E) = 3.$$

We may assume that the quotient $E/\mathcal{F}$ is torsion-free. Then, by Proposition 1.32, it follows $\mathcal{F}$ reflexive. If $\text{rank}(\mathcal{F}) = 1$, then by Proposition 1.33, $\mathcal{F}$ is a line bundle, hence the inclusion $\mathcal{F} \subset E$ gives a non-zero section in $H^0(X, E \otimes \mathcal{F}^*)$, contradiction. If $\text{rank}(\mathcal{F}) = 2$, then $\text{rank}(E/\mathcal{F}) = 1$ and the surjective map $E \to E/\mathcal{F}$ gives an injective map

$$0 \to (E/\mathcal{F})^* \to E^* \,.$$

Since $(E/\mathcal{F})^*$ is reflexive of rank 1 it is a line bundle $L$ and we get a non-zero section in $H^0(X, E^* \otimes L^*)$, contradiction.

**Example 4.26 (J. Coandă)** Let $X$ be as in the Example 4.8 a K3-surface with $\text{Pic}(X) = 0$ $(a(X) = 0)$. We have the exact sequence of holomorphic vector bundles

$$0 \to \mathcal{O}_X \to T_X \otimes T_X \to S^2 T_X \to 0 \ .$$

Since $T_X^* \cong T_X$ and since $T_X$ is simple (otherwise, a non-zero, non-invertible endomorphism $\sigma : T_X \to T_X$ will give a subsheaf $\mathrm{Ker}(\sigma) \subset T_X$ of rank 1, contradicting the irreducibility of $T_X$; see Example 4.8), it follows that

$$H^0(X, T_X \otimes T_X) \cong \mathrm{End}(T_X) \cong \mathbb{C} \ .$$

From $H^1(X, \mathcal{O}_X) = 0$ we get $H^0(X, E) = 0$, where $E := S^2 T_X$. But $E^* \cong E$ since $T_X^* \cong T_X$, hence $H^0(X, E^*) = 0$. As $\mathrm{Pic}(X) = 0$ it follows that $E = S^2 T_X$ is an irreducible vector bundle of rank 3.

**Proposition 4.27** *Let $E$ be an irreducible holomorphic vector bundle over a complex (connected) manifold. Then $E$ is simple.*

*Proof.* Let $f : E \to E$ be a non-zero endomorphism. Since $E$ is irreducible, $\mathrm{Ker}(f) = 0$ and the homomorphism

$$\det(f) : \det(E) \longrightarrow \det(E)$$

is not zero, hence it is an isomorphism. Then $f$ is an isomorphism.

In [E-F] Elencwajg and Forster showed the existence of irreducible rank two vector bundles over 2-dimensional tori $X$ with $\mathrm{NS}(X) = 0$. This was done by comparing the versal deformation of a filtrable rank 2 vector bundle with the space parametrising extensions producing filtrable vector bundles (see Chapter 5). In this way they proved that in general the versal deformation is richer, hence it contains also irreducible vector bundles.

Using the relative Douady space of quotients associated to the versal deformation of a filtrable vector bundle, Bănică and Le Potier [B-L] showed the existence of irreducible vector bundles $E$ in any rank over surfaces with algebraic dimension zero, for a large range of Chern classes $c_1(E), c_2(E)$. They also showed the existence of irreducible rank 2 vector bundles over surfaces with algebraic dimension one and trivial canonical line bundle. As it was remarked by M. Toma (see [To3]) their proof can be extended to any surface $X$ with $a(X) = 1$. By a remark in Section 1.2, it follows that the Chern classes of all these examples of irreducible vector bundles are in the same range as for the filtrable vector bundles.

By the Theorems 4.17 and 4.21, for fixed rank $r$ and fixed first Chern class $c_1 \in \mathrm{NS}(X)$ the only unknown situations are for

$$\Delta \in [0, m(r, c_1)) \ . \tag{4.7}$$

When $m(r, c_1) \neq 0$ this interval is non-empty and the existence problem I reduces to fill this interval with the "candidates", the non-filtrable vector bundles. Thus, in order to solve the existence problem for surfaces we can ask, together with C. Okonek (see [O-V2]), to develop a *technique to study irreducible (and non-filtrable) vector bundles*.

In [To1] M. Toma found examples of topological vector bundles (with $\Delta = 0$) which do not allow filtrable holomorphic structures, but which do allow non-filtrable and even irreducible holomorphic structures.

**Theorem 4.28** *Every complex topological vector bundle $E$ over a 2-dimensional complex torus $X$ having $c_1(E) \in NS(X)$ and $\Delta(E) = 0$ admits holomorphic structures.*

*Proof.* (sketch; cf. [To1]) Firstly, we need the following

**Lemma 4.29** *Let $X$ be a 2-dimensional complex torus, $a \in NS(X)$ and $p$ a prime number such that $p$ divides $\frac{1}{2}a^2$. Then there exist an unramified covering $q : X' \to X$ of degree $p$ and $a' \in NS(X')$ such that $pa' = q^*(a)$.*

The proof of this technical result uses the description of the Neron-Severi group given by the Appell-Humbert Theorem (Theorem 3.9); see [To1]. We shall give only the idea of the proof.

Let $\Gamma$ be a lattice generated by $\gamma_1, \cdots, \gamma_4$ in $\mathbb{C}^2$ with $X \cong \mathbb{C}^2/\Gamma$. Recall the natural isomorphism:

$$NS(X) \cong \{H \mid H \text{ hermitian } 2 \times 2 \text{ matrix with } \operatorname{Im}(\Pi^t H \overline{\Pi}) \in M_4(\mathbb{Z})\}$$

where $\Pi := (\gamma_1, \cdots, \gamma_4)$ is the period matrix of $X$.

For a hermitian matrix $H \in NS(X)$ we denote

$$A := \operatorname{Im}(\Pi^t H \overline{\Pi})$$

and we have

$$A = \begin{pmatrix} A_1 & A_2 \\ -A_2^t & A_3 \end{pmatrix}; \quad A_1 = \begin{pmatrix} 0 & \theta \\ -\theta & 0 \end{pmatrix}, \quad A_2 = \begin{pmatrix} \alpha & \beta \\ \gamma & \delta \end{pmatrix}, \quad A_3 = \begin{pmatrix} 0 & \tau \\ -\tau & 0 \end{pmatrix}.$$

Then, for the self-intersection of an element $a$ in $NS(X)$ represented by $A$ we have the formula:

$$a^2 = 2(\alpha\delta - \beta\gamma - \theta\tau),$$

and the hypothesis of the lemma becomes

$$p \mid (\alpha\delta - \beta\gamma - \theta\tau).$$

We shall take tori $X'$ obtained by factorizing $\mathbb{C}^2$ through lattices obtained by multiplying by $p$ one of $\Gamma$'s generators $\gamma_i$ and preserving the others. The projection $q : X' \to X$ will be an unramified covering of degree $p$. If $\tilde{\Pi}$ is the period matrix thus obtained for $X'$, then we need to get

$$\operatorname{Im}(\tilde{\Pi}^t H \overline{\tilde{\Pi}}) \in M_4(p\mathbb{Z}).$$

The element $H/p \in NS(X')$ would be the looked for $a'$. In order to reach this purpose (i.e. that $A_i \in M_2(p\mathbb{Z})$ for all $i = 1, 2, 3$) we will make a suitable base change for $\Gamma$. Another base of $\Gamma$, $(\gamma_i')$, $i = 1, \cdots, 4$, is related to the previous one by a matrix $M \in M_4(\mathbb{Z})$ with $\det(M) = \pm 1$:

$$\gamma_i' = \sum_j m_{ji} \gamma_j,$$

giving the corresponding period matrix

$$\Pi' = \Pi M .$$

For

$$\mathrm{Im}(\Pi'^t H \overline{\Pi'}) = A' = \begin{pmatrix} A'_1 & A'_2 \\ -A'^t_2 & A'_3 \end{pmatrix},$$

we get

$$A' = M^t A M .$$

Writing

$$M = \begin{pmatrix} a & b \\ c & d \end{pmatrix}$$

with $a$, $b$, $c$, $d \in M_2(\mathbb{Z})$, we get the relations:

$$A'_1 = a^t A_1 a - c^t A^t_2 a + a^t A_2 c + c^t A_3 c ,$$

$$A'_2 = a^t A_1 b - c^t A^t_2 b + c^t A_2 d + c^t A_3 d ,$$

$$A'_3 = b^t A_1 b - d^t A^t_2 b + b^t A_2 d + d^t A_3 d .$$

By a careful analysis of these relations, one can see that we can choose a suitable matrix $M \in M_4(\mathbb{Z})$ with $\det(M) = \pm 1$ and the proof of lemma is over.

Now we shall use induction on the number $n$ of prime factors of $r = \mathrm{rank}(E)$;

$$r = \prod_{i=1}^{k} p_i^{n_i} , \quad n = \sum_{i=1}^{k} n_i .$$

For $n = 0$, i.e. $r = 1$, the statement is true since $c_1(E) \in \mathrm{NS}(X)$ by assumption. Assume the theorem is true for $n$ and we shall prove it for $n + 1$. Let $p$ be a prime factor of $r$. The condition

$$\Delta = \frac{1}{r} \left( c_2(E) - \frac{r-1}{2r} c_1^2(E) \right) = 0$$

becomes

$$r c_2(E) = (r - 1)\frac{1}{2} c_1^2(E) ,$$

and implies $p$ divides $\frac{1}{2} c_1^2(E)$, since the quadratic intersection form on Neron-Severi group $\mathrm{NS}(X)$ is even. By the lemma above, it follows that there exist an unramified covering of degree $p$, $q : X' \to X$ and $a' \in \mathrm{NS}(X')$ such that $pa' = q^*(c_1(E))$, where $X'$ is again a 2-torus. Consider on $X'$ the topological vector bundle $F$ having

$$\mathrm{rank}(F) = \frac{r}{p} , \quad c_1(F) = a' ,$$

$$c_2(F) = \frac{\frac{r}{p} - 1}{2r} c_1^2(E) \in \mathbb{Z} ;$$

note that the map

$$q^* : H^4(X, \mathbb{Z}) \cong \mathbb{Z} \longrightarrow H^4(X', \mathbb{Z}) \cong \mathbb{Z}$$

is given by the multiplication by $p$. A simple computation shows that $\Delta(F) = 0$ and, by induction hypothesis, the topological vector bundle $F$ admits holomorphic structures.

Let $G = \{1, \tau, \cdots, \tau^{p-1}\}$ be the deck-transformations group of the unramified covering $q : X' \to X$ (whose elements are translations of $X'$) and let

$$E' = F \oplus \tau^*(F) \oplus \cdots \oplus (\tau^{p-1})^*(F) \,,$$

be endowed with the holomorphic structure induced by the holomorphic structure of $F$. Then

$$c_1(E') = pc_1(F) = pa' = q^*(c_1(E))$$

$(c_i(\tau^*(F)) = c_i(F)$ since $\tau$ is homotopous to identity map) and

$$c_2 = \frac{p(p-1)}{2}c_1^2(F) + pc_2(F) = \frac{p(r-1)}{2r}c_1^2(E) \,.$$

It follows $\Delta(E') = 0$.

Since $\tau^m$ are translations of $X'$, we get canonical isomorphisms $E' \cong (\tau^m)^*(E')$ compatible with the action of $G$ on $X'$, hence the holomorphic vector bundle $E'$ induces a holomorphic vector bundle $E''$ over $X$ such that $q^*(E'') = E'$. It follows that $\Delta(E'') = 0$ and $c_1(E'') = c_1(E)$, hence the underlying topological vector bundle of $E''$ is $E$.

Let $X$ be a compact complex surface, $a \in NS(X)$ and $r$ a positive integer. We make the following notations:

$$s(r, a) := -\frac{1}{2} \sup_{\mu \in NS(X)} \left( \frac{a}{r} - \mu \right)^2 \,,$$

$$t(r, a) := \inf \left\{ \frac{1}{k(r-k)} s(r, ka) \mid k \in \{1, \cdots, r-1\} \right\} \,.$$

When $X$ is nonalgebraic these numbers are non-negative.

*Remark.* For a nonalgebraic surface $X$ and a filtrable holomorphic vector bundle $E$ of rank $r$ over it, one has

$$\Delta(E) \geq s(r, c_1(E)) \,,$$

by Theorem 4.21 and the inequality $m(r, a) \geq s(r, a)$.

**Lemma 4.30** *Let $X$ be a nonalgebraic surface and $E$ a reducible holomorphic vector bundle of rank $r$ over $X$. Then*

$$\Delta(E) \geq t(r, c_1(E)) \,.$$

*Proof.* Let

$$0 \to E_1 \to E \to E_2 \to 0$$

be an exact sequence with $E_i$ coherent sheaves without torsion of rank $r_i$ and having $c_1(E_i) = a_i$, $i = 1, 2$ ($E$ is reducible). Let $a = c_1(E)$. Then $a_1 + a_2 = a$, $r_1 + r_2 = r$ and, by Riemann-Roch formula for $E$, $E_1$ and $E_2$, we get

$$\Delta(E) = \frac{1}{2r}\left(\frac{a^2}{r} - \frac{a_1^2}{r_1} - \frac{a_2^2}{r_2}\right) + \frac{r_1}{r}\Delta(E_1) + \frac{r_2}{r}\Delta(E_2).$$

Since $\Delta(E_i) \geq 0$ (cf. Theorem 4.17), we have

$$\Delta(E) \geq \frac{1}{2r}\left(\frac{a^2}{r} - \frac{a_1^2}{r_1} - \frac{a_2^2}{r_2}\right) = -\frac{1}{2r_1 r_2}\left(\frac{r_2 a}{r} - a\right)^2 \geq$$

$$\geq \frac{1}{r_2(r - r_2)}s(r, r_2 a) \geq t(r, c_1(E)).$$

We have the following consequences (see [To1]):

**Corollary 4.31** *Let $X$ be a nonalgebraic 2-torus, $r$ a positive integer, $a \in NS(X)$ such that $r$ divides $\frac{1}{2}a^2$ and $r^2$ does not divide $\frac{1}{2}a^2$. Then there exists a topological vector bundle $E$ over $X$ having rank $r$, $c_1(E) = a$ and $\Delta(E) = 0$ admitting holomorphic structures but not filtrable holomorphic structures.*

*Proof.* We choose the topological vector bundle $E$ having

$$c_1(E) = a, \ c_2(E) = \frac{r-1}{2r}a^2.$$

Then $\Delta(E) = 0$ and, by the above theorem, $E$ admits holomorphic structures. Using the remark above it will be enough to verify that in this case $s(r, a) > 0$. Let us assume that $s(r, a) = 0$. Then

$$\sup_{\mu \in NS(X)}\left(\frac{a}{r} - \mu\right)^2 = 0,$$

hence

$$\left(\frac{a}{r} - \mu\right)^2 = 0$$

for some $\mu \in NS(X)$. This implies $a = r\mu + c$ with $c \in NS(X)$ and $c^2 = 0$. Then $c$ is orthogonal on $NS(X)$ (see Corollary 2.10) since $X$ is nonalgebraic. It follows that $a^2 = r^2\mu^2$ and $2r^2$ divides $a^2$, contradiction.

**Corollary 4.32** *If $X$ is a complex 2-torus and $n$ a positive integer such that the group $NS(X)$ is freely generated by an element $a$ with $a^2 = -2n$, then the topological vector bundle $E$ over $X$ of rank $n$ having $c_1(E) = a$ and $\Delta(E) = 0$ admits holomorphic structures but no reducible holomorphic structures.*

*Proof.* It is easy to see in this case that $t(r, a) > 0$ and we apply the above lemma.

Examples of 2-tori as in the above corollaries are easy to construct (see, for example, [To1] or [E-F], Appendix).

*Remark.* In conclusion, the existence problem is still open and it seems to be difficult.

## 4.4 Simple filtrable vector bundles

Simple holomorphic vector bundles admit coarse moduli spaces (see [L-O1, Mj, Nr2]). A natural question which arises is to decide when these moduli spaces are non-empty, hence the question of the existence of simple vector bundles. Unlike irreducible vector bundles which are always simple (see Proposition 4.27), reducible vector bundles may have many endomorphisms in general.

In this section we give the range of Chern classes $c_1, c_2$ of simple filtrable rank two vector bundles over minimal surfaces $X$ with algebraic dimension $a(X) = 0$; (see [To2]), which extends the results for surfaces without divisors (see [B-F2]). This determination is possible in the case $a(X) = 0$ since the quadratic intersection form on Neron-Severi group $NS(X)$ is negative-definite modulo torsion (see [B-F3] or Theorem 2.37).

Recall that a rank 2 holomorphic vector bundle $E$ over the surface $X$ is filtrable if and only if it admits an exact sequence

$$0 \to L_1 \xrightarrow{\alpha} E \xrightarrow{\beta} J_Y \otimes L_2 \to 0 , \qquad (4.8)$$

where $L_1, L_2 \in \operatorname{Pic}(X)$ and $Y$ is a 2-codimensional subspace of $X$ or empty-set, as we have seen in Propositions 4.1 and 4.7. Thus, we have to decide when a filtrable vector bundle given by an extension (4.8) is simple.

**Lemma 4.33** *Under the above hypothesis, we have*

*(1) If $E$ is simple then $H^0(X, L_2^* \otimes L_1) = 0$.*

*(2) If $H^0(X, L_2^* \otimes L_1) = H^0(X, L_1^* \otimes L_2 \otimes J_Y) = 0$ and (4.8) does not split, then $E$ is simple.*

*In particular, if $X$ has no divisors $E$ is simple if and only if $L_1 \not\cong L_2$ and (4.8) does not split.*

*Proof.* (1) For any non-zero section $s \in H^0(X, L_2^* \otimes L_1)$ the composition map

$$E \xrightarrow{\beta} J_Y \otimes L_2 \hookrightarrow L_2 \xrightarrow{\cdot s} L_1 \xrightarrow{\alpha} E$$

gives a non-constant endomorphism of $E$, contradiction.

(2) Assume there exists $\varepsilon \in \operatorname{End}(E)$ a non-zero non-invertible element. Since $\beta \circ \varepsilon \circ \alpha = 0$ we get a commutative diagram

$$
\begin{array}{ccccccccc}
0 & \longrightarrow & L_1 & \xrightarrow{\alpha} & E & \xrightarrow{\beta} & J_Y \otimes L_2 & \longrightarrow & 0 \\
 & & \gamma \downarrow & & \varepsilon \downarrow & & \delta \downarrow & & \\
0 & \longrightarrow & L_1 & \xrightarrow{\alpha} & E & \xrightarrow{\beta} & J_Y \otimes L_2 & \longrightarrow & 0
\end{array}
$$

where $\gamma$ and $\delta$ are homotheties or zero. By using the standard Ker-Coker lemma one finds easily that $\gamma$ and $\delta$ cannot be simultaneously isomorphisms nor simultaneously zero. Then, if $\delta = 0$ and $\gamma \neq 0$, there exists $\psi : E \to L_1$ such that $\alpha \circ \psi = \varepsilon$. Hence $\alpha \circ \gamma = \varepsilon \circ \alpha = \alpha \circ \psi \circ \alpha$, which gives $\gamma = \psi \circ \alpha$ and the sequence (4.8) splits, contradiction. In the same way, the case $\gamma = 0$, $\delta \neq 0$ is excluded.

*Remarks* (1) If $P \in \mathrm{Pic}(X)$, then $E$ is simple if and only if $E' = E \otimes P$ is simple. Recall that

$$c_1(E') = c_1(E) + 2c_1(P) , \quad \Delta(E') = \Delta(E) .$$

(2) Since

$$m(2, a) = -\frac{1}{2} \sup_{\mu \in NS(X)} \left( \frac{a}{2} - \mu \right)^2 ,$$

then

$$m(2, c_1(E')) = m(2, c_1(E)) .$$

(3) It follows that we have to consider only the classes $c_1 + 2\mathrm{NS}(X)$ of $c_1$ modulo $2\mathrm{NS}(X)$ and it will be enough for our purpose to consider extensions (4.8) with $L_2$ trivial:

$$0 \to L \to E \to J_Y \to 0 , \tag{4.9}$$

i.e., a topological rank 2-vector bundle given by $(c_1, \Delta)$ admits simple filtrable holomorphic structures if and only if there exists a simple holomorphic vector bundle given by an extension (4.9) having the same discriminant $\Delta$, and the first Chern class congruent modulo $2\mathrm{NS}(X)$ with $c_1$.

(4) For a 2-vector bundle as in (4.9) we have

$$c_1(E) = c_1(L) , \quad c_2(E) = \ell(Y) .$$

According to the classification Theorem 2.25 a minimal surface $X$ of algebraic dimension $a(X) = 0$ can only belong to one of the classes: 2-tori, class VII surfaces and K3-surfaces.

In case $X$ is a K3-surface and $a \in \mathrm{NS}(X)$ we define:

$$M(a) := \left\{ -\frac{1}{2} \left( \frac{a}{2} - \mu \right)^2 \mid \mu \in \mathrm{NS}(X) \right\}$$

$$m'(a) := \begin{cases} \inf(M(a) \setminus \{m(2, a)\}) & \text{if } M(a) \neq \{m(2, a)\} \\ \infty & \text{if } M(a) = \{m(2, a)\} \end{cases} .$$

We can now state the result (see [To2, B-F2]):

**Theorem 4.34** *If $X$ is a minimal surface with $a(X) = 0$ there exists a simple filtrable holomorphic 2-vector bundle over $X$ having Chern classes $c_1 \in NS(X)$ , $c_2 \in \mathbb{Z}$ if and only if*

$$\Delta(c_1, c_2) \geq m(2, c_1) ,$$

*excepting exactly the following cases:*

*(1) if $X$ is a torus*

$$\Delta(c_1, c_2) = m(2, c_1) = 0 ;$$

*(2) if X is in class VII*

$$\Delta(c_1, c_2) = m(2, c_1) = 0 ,$$

*unless $b_2(X) = 0$, $X$ without divisors and $c_1 \in 2NS(X)$;*

*(3) if X is a K3-surface*

$$m(2, c_1) = 0 \quad and \quad 0 \le \Delta(c_1, c_2) < \sup\{m'(c_1), 2\} ,$$

$$\Delta(c_1, c_2) = m(2, c_1) = \frac{1}{4} ,$$

$$\Delta(c_1, c_2) = m(2, c_1) = \frac{1}{2} .$$

*Proof.* (1) *The case: X a torus.* Let $X$ be a 2-torus of algebraic dimension zero. Then $X$ has no divisors (otherwise, an irreducible curve $C \subset X$ would give by translations an infinite family of irreducible curves on $X$, since $X$ is homogeneous; contradiction to Theorem 2.16). Consider firstly, $c_1 \in NS(X)$ , $c_2 \in \mathbb{Z}$ such that

$$\Delta(c_1, c_2) \ge m(2, c_1) > 0 ,$$

and let $L \in \text{Pic}(X)$ be such that

$$c_1(L) \in c_1 + 2NS(X) ,$$

$$m(2, c_1) = -\frac{1}{2}\left(\frac{1}{2}c_1(L)\right)^2 .$$

Since $L$ is non-trivial ($m(2, c_1) > 0$) and since $X$ has no divisors, we get

$$H^2(X, L) = H^0(X, L^*) = 0 .$$

By Theorem 4.3, it follows that there exist extensions (4.9) with $E$ holomorphic 2-vector bundle having the needed Chern classes if $Y$ is a union of $2(\Delta - m(2, c_1)) \in \mathbb{Z}_+$ distinct simple points. Moreover, by Riemann-Roch formula we have

$$h^1(X, L) = -\frac{1}{2}L^2 = 4m(2, c_1) > 0 ,$$

hence also in the case $Y = \emptyset$ one has non-trivial extensions of type (4.9). By Lemma 4.33 we get that the filtrable holomorphic 2-vector bundle $E$ is simple.

Consider now the case $m(2, c_1) = 0$. By the Theorem 2.37, it follows that $m(2, c_1) = 0$ if and only if $c_1 \in 2NS(X)$. If $\Delta(c_1, c_2) > m(2, c_1) = 0$, then we choose $L \in \text{Pic}(X)$ as above and we have $c_1(L) = 0$. Now, if we take $L \in \text{Pic}_0(X)$ , $L \not\cong \mathcal{O}_X$ we obtain as above

$$H^2(X, L) = H^0(X, L^*) = 0$$

and $\ell(Y) > 0$. Then, by Theorem 4.3 and Lemma 4.33, we get a simple filtrable holomorphic 2-vector bundle $E$ having Chern classes $c_1 \in NS(X)$ and $c_2 \in \mathbb{Z}$.

If $\Delta(c_1, c_2) = m(2, c_1) = 0$, we get that the filtrable holomorphic 2-vector bundles $E$ are given (up to a tensoring with a line bundle) by the extensions of the form

$$0 \to L \to E \to \mathcal{O}_X \to 0$$

($Y = \emptyset$, $L \in \mathrm{Pic}_0(X)$ !). If $L \cong \mathcal{O}_X$, then by Lemma 4.33, $E$ is not simple. If $L \not\cong \mathcal{O}_X$, then the above extensions are classified by $H^1(X, L) = 0$ (by Riemann-Roch), hence the extensions split, i.e. $E \cong L \oplus \mathcal{O}_X$ and $E$ is not simple.

(2) *The case: X in class VII.* Let $X$ be in class VII, $c_1 \in \mathrm{NS}(X)$, $c_2 \in \mathbb{Z}$ and let $\Delta = \Delta(c_1, c_2)$.

*Subcase (a)* If $\Delta > m(2, c_1)$ choose $L_1 \in \mathrm{Pic}(X)$ such that $c_1(L_1) \in c_1 + 2\mathrm{NS}(X)$ and $m(2, c_1) = -\frac{1}{2} \left( \frac{1}{2} c_1(L_1) \right)^2$. Then twist (if necessary) $L_1$ by $L_0 \in \mathrm{Pic}_0(X)$ in order to have for $L = L_1 \otimes L_0$ the conditions

$$H^0(X, L^* \otimes \mathcal{K}_X) = H^0(X, L) = H^0(X, L^*) = 0 .$$

This is possible since $\mathrm{Pic}_0(X) \cong \mathbb{C} \setminus \{0\}$ (see [B-P-V], p. 188) and the elements in $\mathrm{Pic}(X)$ admitting non-trivial sections form a countable subset. As we have seen, we can take $c_1 = c_1(L)$. The assumption $\Delta > m(2, c_1)$ implies $c_2 > 0$. The condition $H^0(X, L^* \otimes \mathcal{K}_X) = 0$ ensures the existence of an extension (4.9) with a filtrable holomorphic 2-vector bundle $E$ with Chern classes $c_1$ and $c_2$, if we choose $Y$ to be the union of $c_2$ distinct simple points. Then, by the Lemma 4.33 we get that $E$ is simple.

*Subcase (b)* If $\Delta = m(2, c_1) > 0$ take $L$ as above and $Y = \emptyset$. We have $\Delta = -\frac{1}{8} c_1^2(L) = m(2, c_1)$ hence, by Riemann-Roch formula, we get

$$h^1(X, L) = 4\Delta + \frac{1}{2} L.\mathcal{K}_X ,$$

$$h^1(X, L^*) = 4\Delta - \frac{1}{2} L.\mathcal{K}_X + h^0(X, L \otimes \mathcal{K}_X) .$$

It is clear that at least one of these numbers is positive, and consider a corresponding non-trivial extension

$$0 \to L \to E \to \mathcal{O}_X \to 0$$

or

$$0 \to L^* \to E \to \mathcal{O}_X \to 0 .$$

Again, by Lemma 4.33, we get that $E$ is a simple filtrable holomorphic 2-vector bundle with given Chern classes.

*Subcase (c)* Let $\Delta = m(2, c_1) = 0$. Firstly, we shall show that if $E$ is a simple filtrable vector bundle of rank 2 having these Chern classes, then necessarily $b_2(X) = 0$, $X$ has no divisors and $c_1 \in 2\mathrm{NS}(X)$. Assume that $E$ has an extension of type (4.9). From hypothesis we get $Y = \emptyset$ and $c_1^2(L) = 0$. Thus $E$ has an extension

$$0 \to L \to E \to \mathcal{O}_X \to 0 ,$$

with $L^2 = 0$ , $h^0(X, L) = 0$ ($E$ is simple !) and $h^0(X, L^* \otimes \mathcal{K}_X) = 1$ (by Corollary 2.10, $c_1^2(L) = 0$ implies $c_1(L).K = 0$ and the Riemann-Roch formula for $L$ will give $h^0(X, L^* \otimes \mathcal{K}_X) = 1$). Thus we get $\mathcal{K}_X \cong L \otimes \mathcal{O}_X(\sum_{i=1}^k r_i C_i)$ with $r_i$ non-negative integers and $C_i$ irreducible curves on $X$. If $\mathcal{K}_X.C_i < 0$ then $C_i^2 < 0$ (otherwise, $C_i^2 = 0$ would imply $\mathcal{K}_X.C_i = 0$ again by Corollary 2.10; but then $C_i$ is exceptional ((-1)-curve) and we get a contradiction with the assumption that $X$ is minimal. It follows that

$$\mathcal{K}_X^2 = \mathcal{K}_X.L + \sum_{i=1}^{k} r_i \mathcal{K}_X.C_i \geq 0 \,.$$

By using Noether formula,

$$\chi(\mathcal{O}_X) = \frac{1}{12}(c_1^2(X) + c_2(X))$$

and the fact that $b_1(X) = 1$ implies $h^1(X, \mathcal{O}_X) = 1$ (see [B-P-V]; Chapter IV 2.6), we get $b_2(X) = -\mathcal{K}_X^2$, hence $b_2(X) = 0$.

If $X$ has no curves it must be an Inoue surface (see [Bo1, L-Y-Z, Te2]) and one can see that in this case $c_1(X) = 0$. Now $L \cong \mathcal{K}_X$ since $h^0(X, L^* \otimes \mathcal{O}_X) = 1$, hence $c_1 = c_1(X) = 0$.

Assume now $X$ has divisors and we shall show that this assumption leads to a contradiction. Since $a(X) = 0$, $b_1(X) = 1$, $b_2(X) = 0$. $X$ must be a Hopf surface by Theorem 2.27. Hence one has the following form for the canonical line bundle

$$\mathcal{K}_X = \begin{cases} \mathcal{O}_X(-C_1 - C_2) & \text{if } X \text{ admits two irreducible curves } C_1,\ C_2 \\ \mathcal{O}_X(-(m+1)C) & \text{if } X \text{ admits only one irreducible curve } C \end{cases} ,$$

where $m$ is a positive integer depending on the transformation group on the universal covering $\mathbb{C}^2 \setminus \{0\}$, giving the surface $X$. Denote $D_1 = \sum_{i=1}^{k} r_i C_i \geq 0$, so that $L \cong \mathcal{K}_X \otimes \mathcal{O}_X(-D_1)$. We can write $\mathcal{K}_X = \mathcal{O}_X(-D_2 - D_3)$ with divisors $D_2 > 0$, $D_3 > 0$ and we have a commutative diagram

$$L \longrightarrow \mathcal{O}_X$$
$$\searrow \quad \swarrow$$
$$\mathcal{O}_X(-D_2)$$

with all the arrows natural inclusions. By Riemann-Roch formula we get the equality $H^1(X, \mathcal{O}_X(-D_2)) = 0$ ($b_2(X) = 0$) and we find that the natural map

$$H^1(X, L) \longrightarrow H^1(X, \mathcal{O}_X)$$

is zero. Since the above extension is non-trivial, the connecting homomorphism $\delta$ in

$$0 \to H^0(X, L) \to H^0(X, E) \to H^0(X, \mathcal{O}_X) \xrightarrow{\delta} H^1(X, L) \to H^1(X, E)$$

is non-trivial. But $H^0(X, \mathcal{O}_X) \cong \mathbb{C}$, hence $H^0(X, E) = H^0(X, L) = 0$. By twisting the above extension by $L^*$ one finds the commutative diagram

$$
\begin{array}{ccccccccc}
0 & \longrightarrow & L & \longrightarrow & E & \longrightarrow & \mathcal{O}_X & \longrightarrow & 0 \\
 & & \downarrow & & \downarrow & & \downarrow & & \\
0 & \longrightarrow & \mathcal{O}_X & \longrightarrow & E^* & \longrightarrow & L^* & \longrightarrow & 0
\end{array}
$$

with vertical arrows given by functoriality and the natural inclusion

$$\mathcal{O}_X \hookrightarrow L^* = \mathcal{O}_X(D_1 + D_2 + D_3) \ .$$

Hence, we have the exact sequence

$$0 \to H^0(X, \mathcal{O}_X) \to H^0(X, E^*) \to H^0(X, L^*) \to H^1(X, \mathcal{O}_X)$$

and the commutative diagram

$$
\begin{array}{ccc}
H^0(X, \mathcal{O}_X) & \xrightarrow{\ \delta\ } & H^1(X, L) \\
\wr\downarrow & & \downarrow \\
H^0(X, L^*) & \longrightarrow & H^1(X, \mathcal{O}_X)
\end{array}
$$

By counting dimensions, we get that $\delta$ and the first vertical map are isomorphisms. Since the second vertical map is zero we obtain

$$h^0(X, E^*) = h^0(X, \mathcal{O}_X) + h^0(X, L^*) = 2 \ .$$

Now, any element in $H^0(X, E^*)$ induces by twisting the above extension with $E^*$ a commutative diagram

$$
\begin{array}{ccccccccc}
0 & \longrightarrow & L & \longrightarrow & E & \longrightarrow & \mathcal{O}_X & \longrightarrow & 0 \\
 & & \downarrow & & \downarrow & & \downarrow & & \\
0 & \longrightarrow & E & \longrightarrow & E \otimes E^* & \longrightarrow & E^* & \longrightarrow & 0
\end{array}
$$

Hence, the diagram

$$
\begin{array}{ccc}
H^0(X, \mathcal{O}_X) & \xrightarrow{\ \delta\ } & H^1(X, L) \\
\downarrow & & \downarrow \\
\end{array}
$$
$$
0 \longrightarrow H^0(X, E \otimes E^*) \longrightarrow H^0(X, E^*) \longrightarrow H^1(X, E)
$$

is commutative. If we prove that the arrow $H^0(X, E^*) \to H^1(X, E)$ is zero, then we find $h^0(X, E \otimes E^*) = 2$, contradicting the assumption that $E$ is simple.

Let $\alpha_1, \alpha_2$ be linearly independent in $H^0(X, E^*)$ with $\alpha_1$ induced by the morphism $L \to E$. By the above facts the morphism

$$H^1(X, L) \xrightarrow{H^1(\alpha_1)} H^1(X, E)$$

is zero. Denote by $\beta_1$ the morphism $E \to \mathcal{O}_X$. Then $\beta_1 \circ \alpha_2 \neq 0$, hence $\beta_1 \circ \alpha_2$ must be proportional to the natural inclusion $L \to \mathcal{O}_X$. We denoted also by $\alpha_2$ the morphism $L \to E$ inducing $\alpha_2 \in H^0(X, E^*)$. Thus $H^1(\beta_1 \circ \alpha_2) = 0$. Since $H^1(\beta_1)$ is injective, we find that

$$H^1(X, L) \xrightarrow{H^1(\alpha_2)} H^1(X, E)$$

is zero. As $\alpha_1, \alpha_2$ generate $H^0(X, E^*)$, it follows that the arrow $H^0(X, E^*) \to H^1(X, E)$ must be zero and $E$ would not be simple, contradiction. Thus, we reduced the proof to the following

*Subcase (d)* When $\Delta = m(2, c_1) = 0$, $X$ in class VII without divisors, $b_2(X) = 0$ and $c_1 \in 2\mathrm{NS}(X)$.

By twisting with a line bundle we can reduce the problem to the extensions of the form

$$0 \to \mathcal{K}_X \to E \to \mathcal{O}_X \to 0$$

($c_1(X) = 0$ !). Now, since by Riemann-Roch $h^1(X, \mathcal{K}_X) = 1$, it follows by the Lemma 4.33, that in this case there exist simple filtrable holomorphic 2-vector bundles.

(3) *The case: X a K3-surface.* For the very nice but very long proof in the case of a K3-surface with divisors see [To2].

We shall give the proof only in the case of a K3-surface without divisors ( [B-F2]).

**Theorem 4.35** *Let X be a K3-surface without divisors. There exists a simple filtrable holomorphic rank 2-vector bundle over X having Chern classes $c_1 \in NS(X)$, $c_2 \in \mathbb{Z}$ if and only if*

$$\Delta(c_1, c_2) \geq m(2, c_1) ,$$

*excepting exactly the following cases:*

$$m(2, c_1) = 0 \quad and \quad 0 \leq \Delta < m'(c_1)$$

$$m(2, c_1) = \Delta(c_1, c_2) = \frac{1}{2} .$$

*Proof.* By the Theorem 2.37 we have $m(2, c_1) = 0$ if and only if $c_1 \in 2\mathrm{NS}(X)$. Now, for any $L \in \mathrm{Pic}(X)$ with $L \not\equiv \mathcal{O}_X$ we have

$$H^0(X, L) = H^0(X, L^*) = 0 ,$$

since $X$ has no divisors. By Riemann-Roch formula we get

$$c_1^2(L) = -4 - 2h^1(X, L) \leq -4 ,$$

hence

$$m(2, c_1) \geq \frac{1}{2}$$

for any $c_1 \notin 2\mathrm{NS}(X)$.

If $\Delta(c_1, c_2) \geq m(2, c_1) > 1/2$, then as in the case of a 2-torus we choose $L \in \mathrm{Pic}(X)$ such that

$$c_1(L) \in c_1 + 2\mathrm{NS}(X) ,$$

$$m(2,c_1) = -\frac{1}{2}\left(\frac{1}{2}c_1(L)\right)^2 .$$

Similarly, since $L$ is non-trivial and since $X$ has no divisors, we get

$$H^2(X,L) = H^0(X,L^*) = 0 .$$

By Theorem 4.3 we obtain extensions (4.9) with filtrable holomorphic 2-vector bundles having the needed Chern classes if we choose $Y$ to be a union of $2(\Delta - m(2,c_1))$ distinct simple points or $Y = \emptyset$ in case $\Delta(c_1,c_2) = m(2,c_1)$. In either case, we have

$$h^1(X,L) = -2 + 4m(2,c_1) \geq 1 ,$$

hence there are non-trivial extensions (4.9), which by Lemma 4.33 are simple.

If $\Delta(c_1,c_2) > m(2,c_1) = 1/2$ we get as above filtrable holomorphic 2-vector bundles $E$ given (up to tensoring by a line bundle) by extensions

$$0 \to L \to E \to J_Y \to 0 ,$$

and again by Lemma 4.33 they are simple.

If $\Delta(c_1,c_2) = m(2,c_1) = 1/2$, we get an extension

$$0 \to L \to E \to \mathcal{O}_X \to 0$$

($Y = \emptyset$ since $\ell(Y) = 2(\Delta - m(2,c_1)) = 0$) with $c_1(E) = c_1(L)$ ($c_1^2(L) = -4$) and $c_2(E) = 0$. By Riemann-Roch formula, it follows

$$h^1(X,L) = -2 + 2 = 0 ,$$

hence $E \cong \mathcal{O}_X \oplus L$ and $E$ is not simple.

If $m(2,c_1) = 0$ then $c_1 \in 2\mathrm{NS}(X)$ and $m'(c_1) = m'(0)$. Let $L \in \mathrm{Pic}(X)$ such that

$$-\frac{1}{2}\left(\frac{1}{2}c_1(L)\right)^2 = m'(0) > 0 .$$

Since $L$ is non-trivial we have again

$$H^0(X,L) = H^0(X,L^*) = 0 .$$

Now, if $\Delta > m'(0)$, then we take $Y$ a union of $2(\Delta - m'(0))$ distinct simple points and, by the Theorem 4.3, we get a filtrable holomorphic 2-vector bundle $E$ as an extension

$$0 \to L \to E \to J_Y \to 0$$

with $c_1(E) = c_1(L) \in 2\mathrm{NS}(X)$, $c_2(E) = \ell(Y) = 2(\Delta - m'(0))$ and $\Delta(E) = \Delta$, which is simple.

For $\Delta = m'(0)$ we have $Y = \emptyset$ and by Riemann-Roch

$$h^1(X,L) = -2 - m'(0)/2 \geq 2 ,$$

etc. Clearly, if $\Delta < m'(0)$ and if we suppose that there exist simple filtrable holomorphic 2-vector bundles with given invariants, then we get

$$\Delta = \frac{1}{2}\ell(Y) - \frac{1}{2}\left(\frac{1}{2}c_1(L)\right)^2 ,$$

which implies $\Delta \geq m'(0)$, contradiction.

# 5. Classification of vector bundles

This chapter is devoted to the classification problem for holomorphic vector bundles over compact complex surfaces. We present without proofs fundamental results in the theory of deformations of vector bundles and we give some applications to filtrable and non-filtrable (even irreducible) vector bundles over nonalgebraic surfaces. A very rough idea about the existence of a coarse moduli space $\mathcal{M}_X^{s,E}$ of simple holomorphic vector bundles in a fixed $\mathcal{C}^\infty$-complex vector bundle $E$ over a compact complex manifold $X$ is given and some examples are presented. The notion of stability of holomorphic vector bundles in its natural (historical) development is described: firstly, for projective manifolds, then for Kähler manifolds and finally, for arbitrary compact complex manifolds (endowed with a Gauduchon metric). General properties of stable vector bundles (or, more generally, of stable torsion-free coherent sheaves) are presented for the convenience of the reader and some examples are given. Then we give only the general idea of the construction of the coarse moduli spaces $\mathcal{M}_X^{g-st,E}$ of stable vector bundles in a fixed $\mathcal{C}^\infty$-complex vector bundle $E$ over a (Gauduchon) compact complex manifold $X$. Finally, we present some moduli spaces of algebraic 2-vector bundles over ruled surfaces, which are constructed without the use of the stability notion.

## 5.1 Deformations of vector bundles and applications

Let $X$ be a compact complex space and let $E$ be a holomorphic vector bundle over $X$. Let $\mathcal{A}n_0$ denote the category of germs of (analytic) complex spaces.

**Definition 5.1** (1) By a *deformation* of $E$ we mean a triple $\xi = (S, \mathcal{E}, \tau)$, where $(S, 0) \in \mathcal{A}n_0$ is a germ of complex space, $\mathcal{E}$ is a holomorphic vector bundle over $S \times X$ and $\tau : E \xrightarrow{\sim} \mathcal{E}|_{\{0\} \times X}$ is an isomorphism of holomorphic vector bundles (over the canonical isomorphism $X \xrightarrow{\sim} \{0\} \times X$).

(2) If $\xi' = (S', \mathcal{E}', \tau')$ is another deformation of $E$, then a *morphism of deformations* $\xi' \to \xi$ is a pair $(\alpha, \varphi)$ such that $\alpha : S' \to S$ is a morphism of germs of complex spaces, $\varphi : \mathcal{E}' \to (\alpha \times id_X)^*(\mathcal{E})$ is a morphism of holomorphic vector bundles, compatible with the isomorphisms $\tau$ and $\tau'$, i.e. the following diagram of isomorphisms is commutative:

$$
\begin{array}{ccc}
E & \xrightarrow[\tau']{\sim} & \mathcal{E}'|_{\{0\}\times X} \\
\wr\downarrow id & & \wr\downarrow \varphi_0 \\
E & \xrightarrow[\tau]{\sim} & \mathcal{E}|_{\{0\}\times X}
\end{array}
$$

Let $\xi = (S, \mathcal{E}, \tau)$ be a deformation of $E$ and let $\alpha : S' \to S$ be a morphism of germs of complex spaces. By taking $\mathcal{E}' := (\alpha \times id_X)^*(\mathcal{E})$, $\varphi = id_{\mathcal{E}'}$, the pair $(\alpha, \varphi) : \xi' \to \xi$ is a morphism of deformations, where $\xi' = (S', \mathcal{E}', \tau')$ with $\tau' : E \xrightarrow{\sim} \mathcal{E}'|_{\{0'\}\times X}$ the natural isomorphism induced by $\tau$. The deformation $\xi'$ is up to an isomorphism uniquely determined and is denoted by $\alpha^*(\xi)$.

**Definition 5.2** Let $S$ be a germ of complex space, $A$ its local ring (i.e. $A := \mathcal{O}_{S,0}$) and $\boldsymbol{m}_0$ the maximal ideal of $A$. The *tangent space* of $S$ is

$$
T(S) := (\boldsymbol{m}_0/\boldsymbol{m}_0^2)^* \cong \operatorname{Der}(A, \mathbb{C}) .
$$

*Remarks.* (1) Every morphism of germs of complex spaces $\alpha : S' \to S$ induces a $\mathbb{C}$-linear map

$$
T(\alpha) : T(S') \longrightarrow T(S) .
$$

(2) If $\alpha : S' \to S$ is a morphism of germs of complex spaces and if $T(\alpha) : T(S') \longrightarrow T(S)$ is injective, then $\alpha$ is an immersion.

(3) If $\alpha : S \to S$ is an endomorphism of a germ of complex space $S$ and if $T(\alpha)$ is bijective, then $\alpha$ is an automorphism.

**Definition 5.3** (1) A deformation $\xi$ of $E$ is called *complete*, if for every deformation $\xi'$ of $E$ there exists a morphism $\xi' \to \xi$.

(2) A deformation $\xi = (S, \mathcal{E}, \tau)$ of $E$ is called *effective*, if for every deformation $\xi' = (S', \mathcal{E}', \tau')$ and any two morphisms $(\alpha, \varphi) : \xi' \to \xi$ and $(\beta, \psi) : \xi' \to \xi$ we have $T(\alpha) = T(\beta)$.

(3) A deformation $\xi$ of $E$ is called *versal (semi-universal)* if it is complete and effective.

*Remark.* By remark (3) above it follows that a versal deformation is unique up to a (non-canonical) isomorphism.

For a proof of the following fundamental result, see [F-K]:

**Theorem 5.4 (Forster-Knorr)** *Let $E$ be a holomorphic vector bundle over a compact complex space $X$. Then there exists a versal deformation of $E$ and the tangent space of the basis $S$ of the versal deformation is $T(S) \cong H^1(X, \mathcal{E}nd(E))$.*

The idea of the proof is the following: by the results of Schlesinger [Sc] and Schuster [Ss1] there exists a formal versal deformation of $E$; using the description of the deformations by cocycles with classes in the cohomology set $H^1(X, GL(n, \mathcal{O}_{S \times X}))$ one can prove by Grauert division theorem and Grauert's extensions method that we obtain a "convergent solution".

In the case of a compact complex manifold $X$, the Kuranishi theory of local deformations for vector bundles can be developed (see, for example, [F-M3], p. 293, [Mj]).

We shall present the examples of irreducible holomorphic vector bundles constructed by Elencwajg-Forster using deformations of filtrable holomorphic vector bundles; see [E-F], which we follow closely. We need some preliminary results.

*Projective $(r-1)$-bundles* (i.e. holomorphic fibre bundles with fibre $\mathbb{P}^{r-1}$ and structure group $PGL(r, \mathbb{C})$) on a complex space $X$ are classified by $H^1(X, PGL(r, \mathcal{O}_X))$ and every holomorphic vector bundle $E$ of rank $r$ over $X$ gives rise to a projective $(r - 1)$- bundle $\mathbb{P}(E)$ (see Section 1.2). The exact sequence

$$0 \to \mathcal{O}_X^* \to GL(r, \mathcal{O}_X) \to PGL(r, \mathcal{O}_X) \to 0$$

gives the associated exact cohomology sequence

$$H^1(X, GL(r, \mathcal{O}_X)) \to H^1(X, PGL(r, \mathcal{O}_X)) \to H^2(X, \mathcal{O}_X^*) .$$

Thus, if $H^2(X, \mathcal{O}_X^*) = 0$ (in particular if $X$ is a curve or $\mathbb{P}^n$) every projective bundle is of the form $\mathbb{P}(E)$, where $E$ is a holomorphic vector bundle. Generally, this is no longer true but, if $P$ is a projective bundle associated to a vector bundle, then any small deformation of $P$ also comes from a vector bundle (see [E-F]):

**Theorem 5.5** *Let $E$ be a holomorphic vector bundle of rank $r$ over a compact complex space $X$. Let $P \to S \times X$ be a deformation of $\mathbb{P}(E)$ over the germ of complex space $(S, 0)$. Then there exists a deformation $\mathcal{E} \to S \times X$ of the vector bundle $E$ such that $P \cong \mathbb{P}(\mathcal{E})$. Moreover, one can choose $\mathcal{E}$ such that $\det(\mathcal{E})$ is a trivial deformation of $\det(E)$ and, with this supplementary condition $\mathcal{E}$ is uniquely determined.*

Given a holomorphic vector bundle $E$ of rank $r$ over a complex space $X$ we have a canonical injective map

$$\mathcal{O}_X \longrightarrow \mathcal{E}nd(E) , \quad f \mapsto f.\mathrm{id}_E .$$

This injection splits by the map

$$\varphi \mapsto \frac{1}{r}\mathrm{trace}(\varphi)$$

and we get a direct sum decomposition

$$\mathcal{E}nd(E) \cong \mathcal{O}_X \oplus sl(E) ,$$

where $sl(E)$ is the sheaf of endomorphisms of trace zero. In particular, we have for any $q \in \mathbb{N}$

$$H^q(X, \mathcal{E}nd(E)) \cong H^q(X, \mathcal{O}_X) \oplus H^q(X, sl(E)) .$$

Consider the projective bundle $\mathbb{P}(E)$ associated to $E$. If $X$ is compact, it is known that the versal deformation of $\mathbb{P}(E)$ exists and the tangent space of the basis of the versal deformation is $H^1(X, sl(E))$. We have the following result (see [E-F]):

**Theorem 5.6** *Let $E$ be a holomorphic vector bundle over the compact complex space $X$. Let $\mathcal{E}' \to \Sigma \times X$ be a deformation of $E$ such that $\mathbb{P}(\mathcal{E}') \to \Sigma \times X$ is the versal deformation of $\mathbb{P}(E)$. Let $\mathcal{L} \to \Pi \times X$ be the versal deformation of the trivial line bundle over $X$. Then the exterior tensor product*

$$\mathcal{L} \boxed{\times} \mathcal{E}' \longrightarrow (\Pi \times \Sigma) \times X$$

*is the versal deformation of $E$.*

*Proof.* (cf. [E-F]) The deformation $\mathcal{E}' \to \Sigma \times X$ exists by Theorem 5.5. The versal deformation $\mathcal{L} \to \Pi \times X$ of the trivial line bundle is obtained as follows: choose cocycles $(h_{ij}^k) \in Z^1(\mathcal{U}, \mathcal{O}_X)$, $k = 1, \cdots, m$ whose cohomology classes form a basis of $H^1(X, \mathcal{O}_X)$. Then $\Pi = (\mathbb{C}^m, 0)$ and the cocycle defining $\mathcal{L}$ is given by

$$g_{ij} := \exp\left(\sum_{k=1}^m t_k h_{ij}^k\right) ,$$

where $t_1, \cdots, t_m$ are the coordinates in $\mathbb{C}^m$.

Let $\mathcal{E} \to S \times X$ be the versal deformation of $E$. Then $\mathbb{P}(\mathcal{E}) \to S \times X$ is a deformation of $\mathbb{P}(E)$, hence there exists a map $\alpha : S \to \Sigma$ such that

$$\mathbb{P}(\mathcal{E}) \cong \alpha^* \mathbb{P}(\mathcal{E}') = \mathbb{P}(\alpha^* \mathcal{E}').$$

Then there exists a deformation $\mathcal{M} \to S \times X$ of the trivial line bundle such that

$$\mathcal{E} \cong \mathcal{M} \otimes \alpha^* \mathcal{E}' .$$

By the versal property of $\mathcal{L} \to \Pi \times X$, there exists a map $\beta : S \to \Pi$ such that $\mathcal{M} \cong \beta^* \mathcal{L}$. Thus, letting

$$f := (\beta, \alpha) : S \to \Pi \times \Sigma ,$$

we have

$$\mathcal{E} \cong f^*(\mathcal{L} \boxed{\times} \mathcal{E}') .$$

On the other hand, by the versal property of $\mathcal{E} \to S \times X$, there exists a map $g : \Pi \times \Sigma \to S$ such that

$$\mathcal{L} \boxed{\times} \mathcal{E}' \cong g^*(\mathcal{E}) .$$

It follows,

$$\mathcal{E} \cong (g \circ f)^*(\mathcal{E}) ,$$

which implies

$$(dg)_0 \circ (df)_0 = d(g \circ f)_0 = id_{T_0(S)} .$$

Consider the sequence

$$T_0(S) \overset{(df)_0}{\to} T_{(0,0)}(\Pi \times \Sigma) \overset{(dg)_0}{\to} T_0(S) .$$

Since

$$T_0(\Pi) = H^1(X, \mathcal{O}_X) , \quad T_0(\Sigma) = H^1(X, sl(E))$$

and

$$T_0(S) = H^1(X, \mathcal{E}nd(E)) ,$$

we have

$$\dim T_0(S) = \dim T_{(0,0)}(\Pi \times \Sigma) \, ,$$

hence $(df)_0$ and $(dg)_0$ are isomorphisms. This implies that $f : S \to \Pi \times \Sigma$ is an isomorphism of germs of complex spaces and

$$\mathcal{L} \boxed{\times} \mathcal{E}' \longrightarrow (\Pi \times \Sigma) \times X$$

is isomorphic to the versal deformation $\mathcal{E} \to S \times X$.

**Corollary 5.7** *Let $E$ be a holomorphic vector bundle over a compact complex space $X$ such that*

$$\dim H^2(X, \mathcal{E}nd(E)) = \dim H^2(X, \mathcal{O}_X) \, .$$

*Then the basis $S$ of the versal deformation of $E$ is smooth.*

*Proof.* The hypothesis implies $H^2(X, sl(E)) = 0$, therefore the basis $\Sigma$ of the versal deformation of $\mathbb{P}(E)$ is smooth (by general deformations arguments). It follows $S = \Pi \times \Sigma$ is smooth and, by the above theorem, $S$ is the basis of the versal deformation of $E$.

**Corollary 5.8** *Let $X$ be a smooth compact complex surface with trivial canonical bundle and let $E$ be a simple vector bundle over $X$. Then the basis of the versal deformation of $E$ is smooth.*

*Proof.* By Serre duality, we have

$$H^2(X, \mathcal{E}nd(E)) \cong H^0(X, \mathcal{E}nd(E)) \cong \mathbb{C} \, ,$$

since $E$ is simple. On the other hand, since $\mathcal{K}_X$ is trivial, we get

$$H^2(X, \mathcal{O}_X) \cong H^0(X, \mathcal{O}_X) \cong \mathbb{C} \, .$$

Then, by the above corollary, the basis of the versal deformation of $E$ is smooth.

Let $X$ be a compact complex (connected) manifold and let $L$, $L'$ be two line bundles over $X$. If $f : L' \to L$ is a non-zero morphism and if $D$ denotes the zero divisor of $f$ ($f : L' \to L$ defines a non-zero morphism $\mathcal{O}_X \to L'^* \otimes L$, etc.), then $L' \cong L \otimes \mathcal{O}_X(D)$, where $\mathcal{O}_X(D)$ is the line bundle associated to $D$. If $X$ is a compact complex manifold without divisors and, if $f : L' \to L$ is a non-zero morphism, then we get $L' \cong L$. We have (see [E-F]):

**Lemma 5.9** *Let $X$ be a compact complex manifold without divisors. Then every filtrable and indecomposable holomorphic 2-vector bundle $E$ over $X$ has a uniquely determined extension*

$$0 \to L \to E \to J_Y \otimes M \to 0 \, ,$$

*where $L$, $M$ are line bundles over $X$ and $Y \subset X$ is a 2-codimensional (or empty) analytic subspace of $X$.*

*Proof.* By Proposition 4.7 $E$ has an extension

$$0 \to L \xrightarrow{\alpha} E \xrightarrow{\beta} J_Y \otimes M \to 0 \,.$$

Suppose $M \not\cong L$ and let $f : L' \to E$ be any monomorphism of a line bundle $L'$ in $E$. Then

$$\beta \circ f : L' \longrightarrow J_Y \otimes M$$

is zero. Indeed, otherwise $Y$ would be empty (since $X$ has no divisors) and $\beta \circ f : L' \to M$ would be an isomorphism. But this would imply $E \cong L \oplus M$, contradicting the indecomposability of $E$. Therefore $f$ factorises as follows

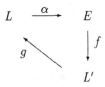

and $g$ is necessarily an isomorphism. This implies the uniqueness of the extension.

If $M \cong L$, denote by $\varepsilon$ the composed morphism

$$E \to J_Y \otimes M \hookrightarrow M \xrightarrow{\sim} L \to E \,.$$

Obviously, $\varepsilon$ is non-zero and $\varepsilon^2 = 0$. Then $E$ is non-simple and the uniqueness of the extension follows from Proposition 4.12.

**Theorem 5.10** *Let $X$ be a compact complex manifold without divisors, $S$ a Stein manifold with $H^2(S, \mathbb{Z}) = 0$ and $E$ a holomorphic vector bundle of rank 2 over $S \times X$. For $s \in S$ denote by $i_s$ the inclusion map*

$$i_s : X \xrightarrow{\sim} \{s\} \times X \hookrightarrow S \times X$$

*and $E_s := i_s^*(E)$. Suppose that $E_s$ is filtrable and indecomposable for all $s \in S$. Then there exist line bundles $L \to S \times X$ , $M \to S \times X$ and a subspace $Y \subset S \times X$ of codimension 2 which is flat over $S$, such that $E$ fits into an exact sequence*

$$0 \to L \to E \to J_Y \otimes M \to 0 \,,$$

*whose restriction to every fibre $\{s\} \times X$ is the uniquely determined extension of $E_s$.*

*Proof.* (cf. [E-F]) Let $\mathcal{P} \to \mathrm{Pic}(X) \times X$ be the Poincaré universal line bundle and consider the vector bundle

$$\mathcal{P}^* \boxtimes E \longrightarrow (\mathrm{Pic}(X) \times S) \times X \,.$$

Let $p : (\mathrm{Pic}(X) \times S) \times X \to \mathrm{Pic}(X) \times S$ be the projection. By the semicontinuity Theorem (see [B-S], p. 134) the set

$$S' := \left\{ (\xi, s) \in \mathrm{Pic}(X) \times S \mid H^0(p^{-1}(\xi, s), \mathcal{P}_\xi^* \boxtimes E_s) \neq 0 \right\}$$

is analytic. Since the extension of every bundle $E_s$ is uniquely determined, the projection $q : S' \to S$ is bijective, hence biholomorphic if we provide $S'$ with the structure of a reduced subspace of $\text{Pic}(X) \times S$. Let

$$\varphi : S \to S' \subset \text{Pic}(X) \times S$$

be the inverse map of $q$ and define the line bundle $L \to S \times X$ by

$$L := (\varphi \times id_X)^* \mathcal{P} \ .$$

For every $s \in S$, the vector space $\text{Hom}(L_s, E_s)$ is one-dimensional, hence the direct image sheaf

$$\pi_* \mathcal{H}om(L, E) \ ,$$

where $\pi : S \times X \to S$ is the projection, is locally free of rank 1 over $S$. The hypothesis $H^2(S, \mathbb{Z}) = 0$ implies

$$\pi_*(\mathcal{H}om(L, E)) \cong \mathcal{O}_S \ .$$

Let $\alpha : L \to E$ be the morphism corresponding to a global non-vanishing section of $\pi_*(\mathcal{H}om(L, E))$. The restriction $\alpha_s : L_s \to E_s$ of $\alpha$ to any fibre $\pi^{-1}(s)$ is, up to a constant factor, the unique monomorphism of a line bundle into $E_s$. The image $\alpha(L)$ is a direct summand of $E$ outside a set of codimension 2. The quotient $E/\alpha(L)$ is torsion-free. Indeed, take the subsheaf $\widehat{\alpha(L)} \subset E$

$$\widehat{\alpha(L)} := t^{-1}(\text{Tors}(E/\alpha(L))) \ ,$$

where $t : E \to E/\alpha(L)$ is the quotient map (see Section 1.4). Since $\alpha(L)$ is isomorphic to $L$ and since $\widehat{\alpha(L)}$ is locally free of rank 1 by Propositions 1.32 and 1.33, then the inclusion map $\alpha(L) \hookrightarrow \widehat{\alpha(L)}$ may be considered as a section of the line bundle $\alpha(L)^* \otimes \widehat{\alpha(L)}$, hence $\text{Supp}(\widehat{\alpha(L)}/\alpha(L))$ has pure codimension 1. But $\text{Tors}(E/\alpha(L)) \cong \widehat{\alpha(L)}/\alpha(L)$, contradiction. Now, define the line bundle $M \to S \times X$ by

$$M := (E/\alpha(L))^{**} \ .$$

Then $E/\alpha(L) \cong J_Y \otimes M$ for a certain 2-codimensional subspace $Y \subset S \times X$. Since $Y$ is locally a complete intersection whose intersection with every fibre $\pi^{-1}(s)$ is 2-codimensional, $Y$ is flat over $S$. The morphism $\alpha : L \to E$ together with the quotient map $E \to E/\alpha(L) \cong J_Y \otimes M$ gives the desired exact sequence

$$0 \to L \to E \to J_Y \otimes M \to 0 \ .$$

*Remark.* If $E \to S \times X$ is a vector bundle as in Theorem 5.10 and if

$$0 \to L_s \to E_s \to J_{Y_s} \otimes M_s \to 0$$

is the unique extension of $E_s$, then

$$s \to [L_s] \quad , \quad s \to [M_s]$$

define holomorphic maps $S \to \text{Pic}(X)$. Moreover, there is a holomorphic map

$$S \to D(X), \quad s \mapsto Y_s,$$

where $D(X)$ denotes the Douady space of all compact analytic subspaces of $X$ (see [Du]).

**Theorem 5.11** *On a 2-torus $X$ with $NS(X) = 0$ there exist non-filtrable holomorphic vector bundles $E$ of rank 2 with $c_2(E) = 2$.*

*Proof.* (cf. [E-F]) Since $NS(X) = 0$ we have $c_1(L) = 0$ for every holomorphic line bundle $L$ over $X$. It follows that $m(2, c_1) = 0$. By Theorem 4.34, there exists a simple filtrable holomorphic rank 2-vector bundle $E$ over $X$ having Chern classes $c_1(E) = 0$ and $c_2(E) = 2$, given by an extension

$$0 \to L \to E \to J_Y \otimes M \to 0,$$

where $L, M$ are holomorphic line bundles with $L \ncong M$ and $Y \subset X$ a locally complete intersection of codimension 2 of length $\ell(Y) = 2$, consisting of two simple points $(c_1(L) = c_1(M) = 0)$. By Corollary 5.8, the basis $(V, 0)$ of the versal deformation $\mathcal{E} \to V \times X$ of $E$ is smooth. The dimension of $V$ equals $h^1(X, \mathcal{E}nd(E))$ and can be calculated by Riemann-Roch formula. We have $\chi(\mathcal{O}_X) = 0$ ($X$ is a 2-torus) and $c_1(E) = 0$, hence

$$h^1(X, \mathcal{E}nd(E)) = h^0(X, \mathcal{E}nd(E)) + h^2(X, \mathcal{E}nd(E)) + 4c_2(E) = 10,$$

because

$$h^0(X, \mathcal{E}nd(E)) = h^2(X, \mathcal{E}nd(E)) = 1$$

($E$ is simple and $\mathcal{K}_X = \mathcal{O}_X$).

Since small deformations of simple bundles are simple and have the same Chern classes, this dimension is invariant under small deformations. This implies that the versal deformation of $E$ is also versal in neighbouring points.

Suppose now that all vector bundles $E_s$, $s \in V$, are filtrable. Since $X$ has no divisors ($NS(X) = 0$ implies $a(X) = 0$, see [E-F], Appendix)), it follows that every $E_s$ has a unique extension (by Lemma 5.9). By Theorem 5.10, there exist deformations $\mathcal{L} \to V \times X$ and $\mathcal{M} \to V \times X$ of $L$, resp. $M$ and a two-codimensional subspace $\mathcal{Y} \subset V \times X$, flat over $V$, such that $\mathcal{E}$ fits into an exact sequence

$$0 \to \mathcal{L} \to \mathcal{E} \to \mathcal{J}_\mathcal{Y} \otimes \mathcal{M} \to 0.$$

Since $\mathcal{Y}_0 = Y$ consists of two simple points, also $\mathcal{Y}_s$ consists of two simple points for $s$ sufficiently near 0. We can define a holomorphic map

$$\varphi : V \longrightarrow \text{Pic}_0(X) \times \text{Pic}_0(X) \times S^2(X)$$

by

$$s \mapsto (\mathcal{L}_s, \mathcal{M}_s, \mathcal{Y}_s).$$

Since

$$\dim(\text{Pic}_0(X) \times \text{Pic}_0(X) \times S^2(X)) = 8,$$

then $S := \varphi^{-1}(L, M, Y)$ is a subgerm of complex space of $V$ of dimension $\geq 2$ and we get a family

$$0 \to q^*(L) \to \mathcal{E}|_{S \times X} \to \mathcal{J}_{S \times Y} \otimes q^*(M) \to 0 \,,$$

where $q : S \times X \to X$ is the projection. This family of extensions defines a holomorphic map

$$\psi : S \longrightarrow \operatorname{Ext}^1(J_Y \otimes M, L) \,.$$

The extensions of $J_Y \otimes M$ by $L$ are classified by the group $\operatorname{Ext}^1(J_Y \otimes M, L)$ and we have the exact sequence

$$0 \to H^1(X, M^* \otimes L) \to \operatorname{Ext}^1(J_Y \otimes M, L) \to$$

$$\to H^0(X, \mathcal{E}xt^1(J_Y \otimes M, L)) \to H^2(X, M^* \otimes L).$$

Since $L \not\cong M$ (Theorem 4.34), then $M^* \otimes L \not\cong \mathcal{O}_X$ and we have by Serre duality

$$H^2(X, M^* \otimes L) = H^0(X, M^* \otimes L) = 0 \,,$$

hence by Riemann-Roch $H^1(X, M^* \otimes L) = 0$. Since $Y$ is a locally complete intersection consisting of two simple points we get

$$\mathcal{E}xt^1(J_Y \otimes M, L) \cong \mathcal{O}_Y$$

and

$$\operatorname{Ext}^1(J_Y \otimes M, L) \cong H^0(X, \mathcal{O}_Y) \cong \mathbb{C}^2 \,.$$

By Serre method (Theorem 4.3), the sheaf corresponding to an extension $\xi \in \operatorname{Ext}^1(J_Y \otimes M, L)$ is locally free if and only if its image in $\mathbb{C}^2$ under the above isomorphism has both coordinates different from zero. Two extensions $\xi_1$, $\xi_2$ which differ only by a constant factor $\lambda \in \mathbb{C}^*$ give rise to isomorphic sheaves.

Since $0 \notin \psi(S)$, we have an associated map to $\psi$

$$\overline{\psi} : S \longrightarrow \mathbb{P}(\operatorname{Ext}^1(J_Y \otimes M, L)) \cong \mathbb{P}^1 \,.$$

If $\overline{\psi}(s) = \overline{\psi}(s')$, then $E_s \cong E_{s'}$. Since dim $S \geq 2$, the fibres of $\overline{\psi}$ have dimensions $\geq 1$. Thus there exists a 1-dimensional subgerm of complex space $C \subset S$, such that $\mathcal{E}|_{C \times X}$ is a trivial deformation of $E$. But this is a contradiction to the versality of the deformation $\mathcal{E} \to V \times X$. Hence, there must exist non-filtrable (i.e. irreducible since $\operatorname{rk} = 2$) vector bundles $E_s$ in this deformation.

As we noticed already, by using the relative Douady space of quotients associated to the versal deformation of filtrable holomorphic vector bundles, Bănică and Le Potier showed the existence of many irreducible vector bundles (see [B-L]). Let $X$ be a nonalgebraic surface and recall the following notation for $\mu \in \operatorname{NS}(X) \otimes \mathbb{Q}$:

$$s(\mu) := -\frac{1}{2} \sup_{\xi \in NS(X)} (\mu - \xi)^2 \,;$$

denote $\chi = \chi(\mathcal{O}_X)$. For proofs of the following results see [B-L]:

**Theorem 5.12 (Bănică-Le Potier)** *Let $X$ be a surface with algebraic dimension $a(X) = 0$, $r$ an integer $\geq 2$, $c_1 \in NS(X)$, $c_2 \in \mathbb{Z}$,*

$$\mu = \frac{c_1}{r} \ , \ \ \Delta = \frac{1}{r}\left(c_2 - \frac{r-1}{2r}c_1^2\right) \ .$$

*If*

$$\Delta \geq m(r,c_1) \ , \quad \Delta + K^2/8 > r(1 + \chi/8) \ ,$$

*then there exists an irreducible holomorphic vector bundle of rank $r$, with Chern classes $c_1$ and $c_2$, over $X$.*

**Theorem 5.13** *Let $X$ be a surface with algebraic dimension $a(X) = 1$, $c_1 \in NS(X)$, $c_2 \in \mathbb{Z}$, $\mu = c_1/2$, $\Delta = (c_2 - c_1^2/4)/2$. If*

$$\Delta \geq s(\mu) \ , \quad \Delta + s(\mu) > 1 + \chi/2 \ ,$$

*then there exists an irreducible holomorphic 2-vector bundle over $X$, with Chern classes $c_1$ and $c_2$.*

In fact, in [B-L] this result is proved in the case of a surface with trivial canonical bundle. M. Toma observed that the proof can actually be extended to any surface $X$ with $a(X) = 1$, making use of the following result (see [To3]):

**Proposition 5.14** *Let $X$ be a minimal nonalgebraic surface. Then there is a constant $c$ depending only on $X$, such that for any rank 1 torsion-free coherent sheaf $\mathcal{F}$ over $X$ one has*

$$h^0(X, \mathcal{F} \otimes \mathcal{K}_X) - h^0(X, \mathcal{F}) \leq c \ .$$

## 5.2 Moduli spaces of simple vector bundles

Let $X$ be a compact complex manifold and let $E$ be a $\mathcal{C}^\infty$-complex vector bundle over $X$ with $c_1(E) \in NS(X)$. Let $\bigwedge^{p,q} T_c^*$ be the bundle of complex $(p,q)$-forms and let $(\bigwedge^{p,q} T_c^*) \otimes E$ be the bundle of $(p,q)$-forms with values in $E$. We denote by $\Omega^{p,q}(E)$ the space of sections of this bundle over $X$.

Recall that a holomorphic vector bundle $F$ over $X$ is simple if

$$End(F) = H^0(X, \mathcal{E}nd(F)) = \mathbb{C}.id_F \ .$$

Let $\mathcal{M}_X^{s,E}$ be the set of isomorphism classes of simple holomorphic structures in $E$ (see Section 1.5).

A holomorphic vector bundle $F$ over $X$ differentiably equivalent to $E$, induces a canonical differential operator $((0,1)$-*type semi-connection*)

$$\bar{\partial}_F : \Omega^{0,0}(E) \longrightarrow \Omega^{0,1}(E)$$

satisfying the Leibnitz-rule

$$\bar{\partial}_F(fs) = \bar{\partial}(f) \otimes s + f\bar{\partial}_F(s)$$

for all $f \in \bigwedge^{0,0} T_c^* = \mathcal{C}^\infty(X, \mathbb{C})$, $s \in \Omega^{0,0}(E)$. There exists a natural extension

$$\bar{\partial}_F : \Omega^{p,q}(E) \longrightarrow \Omega^{p,q+1}(E)$$

defined by

$$\overline{\partial}_F(\alpha \otimes s) = \overline{\partial}(\alpha) \otimes s + (-1)^{p+q}\alpha \wedge \overline{\partial}_F(s)$$

for all $\alpha \in \bigwedge^{p,q} T_c^*$, $s \in \Omega^{0,0}(E)$. Since $F$ is holomorphic we get $\overline{\partial}_F^2 = 0$ (i.e. $\overline{\partial}_F$ is *integrable*).

Conversely, let $\overline{\partial}_A$ be a semi-connection in $E$ (of (0,1)-type), i.e. a differential operator

$$\overline{\partial}_A : \Omega^{0,0}(E) \longrightarrow \Omega^{0,1}(E) \,,$$

which satisfies the Leibnitz-rule. If $\overline{\partial}_A$ is integrable in the sense that $\overline{\partial}_A^2 = 0$ (with natural extension as above), then there exists a holomorphic vector bundle $E_A$, $C^\infty$-equivalent to $E$, such that $\overline{\partial}_{E_A} = \overline{\partial}_A$ (this is the vector bundle version of the Newlander-Nirenberg Theorem; see, for example [F-M3], p. 282). By $\overline{\mathcal{D}}(E)$ we denote the set of all semi-connections in $E$. If $\mathcal{E}nd_{diff}(E)$ denotes the vector bundle of differentiable endomorphisms of $E$, then $\overline{\mathcal{D}}(E)$ is an affine space

$$\overline{\mathcal{D}}(E) = \overline{\partial}_A + \Omega^{0,1}(\mathcal{E}nd_{diff}(E)) \,,$$

where $\overline{\partial}_A$ is an arbitrary semi-connection. Denote by

$$\mathcal{H}(E) := \left\{ \overline{\partial}_A \in \overline{\mathcal{D}}(E) \mid \overline{\partial}_A^2 = 0 \right\}$$

the subset of holomorphic structures. Let $\mathrm{Aut}_{diff}(E)$ be the group of differentiable automorphisms of $E$. Two holomorphic structures $\overline{\partial}_A$, $\overline{\partial}_{A'}$ define holomorphically equivalent vector bundles $E_A$, $E_{A'}$ if and only if there exists an $f \in \mathrm{Aut}_{diff}(E)$ such that

$$\overline{\partial}_{A'} \circ f = f \circ \overline{\partial}_A \,.$$

Every semi-connection $\overline{\partial}_A \in \overline{\mathcal{D}}(E)$ induces a semi-connection in

$$\Omega^{p,q}(\mathcal{E}nd_{diff}(E))$$

by

$$\overline{D}_A(\alpha) = \left[\overline{\partial}_A, \alpha\right] = \overline{\partial}_A \circ \alpha + (-1)^{p+q+1}\alpha \circ \overline{\partial}_A$$

for all $\alpha \in \Omega^{p,q}(\mathcal{E}nd_{diff}(E))$. The group $\mathrm{Aut}_{diff}(E)$ of differentiable automorphisms of $E$ acts on $\overline{\mathcal{D}}(E)$ by

$$\overline{\partial}_A \cdot g := g^{-1} \circ \overline{\partial}_A \circ g = \overline{\partial}_A + g^{-1} \circ \overline{D}_A(g) \,.$$

Since the subgroup $\mathbb{C}^*$ of constant automorphisms acts trivially, we can regard this action as an action of the group

$$\mathrm{Aut}_{diff}^* := \mathrm{Aut}_{diff}(E)/\mathbb{C}^*.id_E \,.$$

Consider now the subset of $\mathcal{H}(E)$

$$\mathcal{S}(E) := \{\overline{\partial}_A \in \mathcal{H}(E) \mid H^0(X, \mathcal{E}nd(E_A)) = \mathbb{C}.id_E\}$$

of simple holomorphic structures in $E$. The natural action of the group $\mathrm{Aut}_{diff}^*$ on $\mathcal{H}(E)$ leaves $\mathcal{S}(E)$ invariant and it restrict to a free action on this set. The orbit space is

$$\mathcal{M}_X^{s,E} := \mathcal{S}(E)/\mathrm{Aut}_{diff}^*(E) \ ,$$

the moduli space of simple holomorphic structures in $E$. We have the following basic result (see [L-O1, Mj]):

**Theorem 5.15** *The set of isomorphism classes of simple holomorphic structures in a fixed $C^\infty$ vector bundle $E$ over a compact complex manifold $X$ has a natural structure of a locally Hausdorff complex space of finite dimension. This complex space is a coarse moduli space for simple vector bundles over $X$ with $C^\infty$ support $E$.*

For this subject see also [Nr2, K-O, F-M3, Ko3, O-V4, Bl2].

The natural complex structure on $\mathcal{M}_X^{s,E}$ is in general not reduced and globally not Hausdorff. Two points $[E_A]$, $[E_{A'}]$ represented by $E_A$, $E_{A'}$ can not be separated if and only if there exist non-zero morphisms $\varphi : E_A \to E_{A'}$, $\psi : E_{A'} \to E_A$ such that $\varphi \circ \psi = 0$ and $\psi \circ \varphi = 0$ (see [L-O1, Nr1]).

Local models of the space $\mathcal{M}_X^{s,E}$ can be described as zero sets of germs of Kuranishi maps. This is a general "philosophy" which goes back to Nijenhuis-Richardson paper [N-R].

In order to give an example of moduli space of simple holomorphic structures we present some results of Bănică-Le Potier ( [B-L]).

**Lemma 5.16** *Let $\mathcal{F}$ be an irreducible sheaf of rank $r > 1$ with slope $\mu$ and discriminant $\Delta$ over a nonalgebraic 2-torus $X$. Then $r(\Delta + s(\mu))$ is an integer $\geq 1$, and $> 1$ with the exception $\Delta = 0$ and $s(\mu) = 1/r$.*

*Proof.* (cf. [B-L]) By tensoring $\mathcal{F}$ with a line bundle we can suppose that $s(\mu) = -\mu^2/2$. Then $r(\Delta + s(\mu)) = c_2 - c_1^2/2$ is an integer $\geq 0$. For all $L \in \mathrm{Pic}_0(X)$ we have

$$\mathrm{Hom}(L, \mathcal{F}) = \mathrm{Ext}^2(L, \mathcal{F}) = 0 \ .$$

Let $a \in X$ be a fixed point and let $\mathcal{P}$ be the unique Poincaré bundle over $\mathrm{Pic}_0(X) \times X$ which is trivial on $\mathrm{Pic}_0(X) \times \{a\}$. Then the sheaf

$$\mathcal{R} = R^1 pr_{1*}(\mathcal{H}om(\mathcal{P}, \mathcal{F}))$$

is a vector bundle. In the Grothendieck group of $\mathrm{Pic}_0(X)$, the element

$$pr_{1!}(-\mathcal{H}om(\mathcal{P}, \mathcal{F}))$$

is represented by $\mathcal{R}$, and its Chern character is given by Grothendieck-Riemann-Roch formula (see Section 1.2):

$$ch(pr_{1!}(\mathcal{H}om(\mathcal{P}, \mathcal{F}))) = pr_{1*}(ch(\mathcal{H}om(\mathcal{P}, \mathcal{F}))) \ ,$$

$$ch(\mathcal{H}om(\mathcal{P}, \mathcal{F})) = e^{-c_1(\mathcal{P})}(r + c_1 + c_1^2/2 - c_2) \ ,$$

where $c_1$ and $c_2$ are the Chern classes of $\mathcal{F}$. From the construction of the Poincaré bundle ( [G-H], p. 328) we know that the Chern class

$$c_1(\mathcal{P}) \in H^2(\mathrm{Pic}_0(X) \times X, \mathbb{Z})$$

belongs to

$$H^1(\mathrm{Pic}_0(X), \mathbb{Z}) \otimes H^1(X, \mathbb{Z}) \, .$$

By the identification

$$H^1(\mathrm{Pic}_0(X), \mathbb{Z}) \cong \mathrm{Hom}(H^1(X, \mathbb{Z}), \mathbb{Z}) \, ,$$

the Chern class $c_1(\mathcal{P})$ corresponds to the identity of $H^1(X, \mathbb{Z})$. There exists a basis $(\theta_i)_{i=1,\cdots,4}$ of $H^1(X, \mathbb{Z})$ such that

$$\theta_1 \wedge \theta_2 \wedge \theta_3 \wedge \theta_4$$

is the fundamental class of $X$. Let $(\theta^i)$ be the dual basis. Then

$$c_1(\mathcal{P}) = \sum_i \theta^i \otimes \theta_i \, ,$$

and

$$pr_{1*}\left(\frac{c_1(\mathcal{P})^4}{4!}\right) = \omega \, ,$$

where $\omega$ is the fundamental class of $\mathrm{Pic}_0(X)$. It follows

$$pr_{1*}(ch(\mathcal{H}om(\mathcal{P}, \mathcal{F}))) = c_1^2/2 - c_2 + c_1 + r\omega \, ,$$

where $c_1$ is considered (by Poincaré duality) as an element of

$$\mathrm{Hom}(H^2(X, \mathbb{Z}), \mathbb{Z}) \cong H^2(\mathrm{Pic}_0(X), \mathbb{Z}) \, .$$

The rank $r'$ and the Chern classes $c_1'$, $c_2'$ of the sheaf $\mathcal{R}$ are given by

$$r' = c_2 - c_1^2/2 \, , \; c_1' = -c_1 \, , \; c_2' - c_1'^2/2 = r \, .$$

It follows $r' \neq 0$ (see Theorem 4.17). The discriminant $\Delta'$ of $\mathcal{R}$ is given by the formula

$$\Delta' = \frac{\Delta}{(\Delta + s(\mu))^2} \, ,$$

and $\Delta' = 0$ if and only if $\Delta = 0$. Therefore, $r' > 1$ with exception $\Delta = 0$ and $s(\mu) = 1/r$.

**Theorem 5.17** Let $X$ be a nonalgebraic 2-torus, $\mathcal{F}$ a torsion-free coherent sheaf of rank $r$, with discriminant $\Delta$ and slope $\mu$ over $X$. Suppose $s(\mu) = 0$ or $1/r^2$ and $\Delta < 2/r - s(\mu)$. Then $\mathcal{F}$ is filtrable.

*Proof.* (cf. [B-L]) Let $(\mathcal{F}_i)$ be a filtration of $\mathcal{F}$ with coherent subsheaves $\mathcal{F}_i$ such that $gr_i := \mathcal{F}_i/\mathcal{F}_{i-1}$ are torsion-free and irreducible (i.e. $(\mathcal{F}_i)$ is a maximal filtration). By tensoring $\mathcal{F}$ with a line bundle, we can suppose that $s(\mu) = -\mu^2/2$. If $r_i$, $\mu_i$, $\Delta_i$ denote the rank, the slope and the discriminant of $gr_i$, then we have

$$\sum_i r_i \mu_i = r\mu$$

$$r(\Delta - \mu^2/2) = \sum_i r_i(\Delta_i - \mu_i^2/2) \, ,$$

by the additivity of the Chern character

$$ch_2(\mathcal{F}) = c_1^2/2 - c_2 = -r(\Delta - \mu^2/2) \, .$$

The hypothesis implies $r(\Delta - \mu^2/2) = 0$ or 1. By the Lemma 5.16, if $gr_i$ is of rank $r_i > 1$, then the term $r_i(\Delta_i - \mu_i^2/2)$ is an integer $\geq 1$. It follows that there exists at most one such term $gr_j$ and we have $\Delta_i = 0$ for all $i$, $\mu_i^2 = 0$ for $i \neq j$, and $r_j\mu_j^2 = -2$. For $i \neq j$, the classes $\mu_i$ are isotropic, hence they are in the kernel of the intersection form (Corollary 2.10). It follows

$$\mu^2 = \left(\frac{1}{r}\sum_i r_i\mu_i\right)^2 = (r_j\mu_j/r)^2 \, ,$$

$$s(\mu) = -\mu^2/2 = r_j/r^2 \, .$$

This contradicts the hypothesis $s(\mu) \leq 1/r^2$, hence $r_i = 1$ for all $i$ and $\mathcal{F}$ is filtrable.

Now, we shall present an example of moduli space of simple holomorphic vector bundles (see, also [B-F1]):

**Corollary 5.18** *Let $X$ be a 2-torus with $NS(X) = 0$ and let $E$ be the topological $(C^\infty)$ 2-vector bundle with Chern classes $c_1(E) = 0$, $c_2(E) = 1$. Then, the moduli space $M_X^{s,E}$ of simple holomorphic structures in $E$ is a complex manifold of dimension 6.*

*Proof.* The hypothesis $NS(X) = 0$ implies $c_1(L) = 0$ for any holomorphic line bundle $L \in \operatorname{Pic}(X)$ ($\operatorname{Pic}(X) \cong \operatorname{Pic}_0(X)$). By the Theorem 5.17, we have that every holomorphic 2-vector bundle with Chern classes $c_1 = 0$, $c_2 = 1$ is filtrable. Now, by the Theorem 4.34, we know that the simple filtrable holomorphic 2-vector bundles with Chern classes $c_1 = 0$, $c_2 = 1$ are given by extensions

$$0 \to L \to F \to J_Y \otimes M \to 0 \, ,$$

where $M$, $L$ are holomorphic line bundles with $M \not\cong L$ and $Y$ is a locally complete intersection of codimension 2 in $X$ and length $\ell(Y) = c_2 = 1$ (i.e. $Y$ is a simple point on $X$). Moreover, by Lemma 5.9, it follows that this extension is uniquely determined by $F$.

From the Ext-spectral sequence we get the exact sequence:

$$H^1(X, L \otimes M^*) \to \operatorname{Ext}^1(J_Y \otimes M, L) \to H^0(Y, \mathcal{O}_Y) \to H^2(X, L \otimes M^*) \, .$$

Since $L \otimes M^* \not\cong \mathcal{O}_X$, we have

$$H^0(X, L \otimes M^*) = H^2(X, L \otimes M^*) = 0 \, ,$$

hence $H^1(X, L \otimes M^*) = 0$ (by Riemann-Roch formula). It follows

$$\dim \operatorname{Ext}^1(J_Y \otimes M, L) = c_2 = 1 \, .$$

Consider the complex manifold of dimension 6 ($\operatorname{Pic}_0(X)$ is a 2-torus !)

$$S := \mathrm{Pic}_0(X) \times \mathrm{Pic}_0(X) \times X \ .$$

Since any non-zero element in $\mathrm{Ext}^1(J_Y \otimes M, L)$ gives a simple filtrable holomorphic 2-vector bundle and, since any two non-zero elements in $\mathrm{Ext}^1(J_Y \otimes M, L)$ give isomorphic simple filtrable holomorphic 2-vector bundles with Chern classes $c_1 = 0$, $c_2 = 1$, we get that the moduli space $\mathcal{M}_X^{s,E}$ is isomorphic to an open subset of the 6-dimensional complex manifold $S$.

*Remark.* See [L-O2, Vu4] for other examples of moduli spaces $\mathcal{M}_X^{s,E}$.

## 5.3 Stable vector bundles

Let $X$ be a projective manifold of dimension $n$ polarised by a very ample divisor $H$. Let $\mathcal{E}$ be a torsion-free coherent sheaf of rank $r$ over $X$. Then the number $c_1(\mathcal{E}).H^{n-1}$ is called the *degree of $\mathcal{E}$ relative to $H$* and is denoted by $\deg_H(\mathcal{E})$. The number $\deg_H(\mathcal{E})/r$ is called the *slope of $\mathcal{E}$ relative to $H$* and it is denoted by $\mu_H(\mathcal{E})$.

**Definition 5.19 (Mumford-Takemoto)** A torsion-free coherent sheaf $\mathcal{E}$ over $X$ is *H-stable* if for every coherent subsheaf $\mathcal{F} \subset \mathcal{E}$ with $0 < \mathrm{rk}(\mathcal{F}) < \mathrm{rk}(\mathcal{E})$, we have

$$\mu_H(\mathcal{F}) < \mu_H(\mathcal{E}) \ .$$

A torsion-free coherent sheaf $\mathcal{E}$ over $X$ is *H-semistable* if for every coherent subsheaf $\mathcal{F} \subset \mathcal{E}$ with $0 < \mathrm{rk}(\mathcal{F}) < \mathrm{rk}(\mathcal{E})$, we have

$$\mu_H(\mathcal{F}) \leq \mu_H(\mathcal{E}) \ .$$

Let $X$ be a compact Kähler manifold of dimension $n$ with Kähler metric $g$ and let $\omega$ be its Kähler form; it is a real positive closed (1,1)-form on $X$. Let $\mathcal{E}$ be a torsion-free coherent sheaf over $X$ and let $c_1(\mathcal{E})$ be its real first Chern class; it is represented by a real closed (1,1)-form on $X$. We can define the *degree of $\mathcal{E}$ relative to $g$* by

$$\deg_g(\mathcal{E}) := \int_X c_1(\mathcal{E}) \wedge \omega^{n-1} \ ,$$

and the *slope of $\mathcal{E}$ relative to $g$* by

$$\mu_g(\mathcal{E}) := \deg_g(\mathcal{E})/\mathrm{rk}(\mathcal{E}) \ .$$

Obviously, one can extend the definition of $g$-stability (resp. $g$- semistability) to this case as in Definition 5.19 .

More generally, let $X$ be now a compact complex hermitian manifold of dimension $n$.

**Definition 5.20** A hermitian metric $g$ on $X$ is called *Gauduchon metric* if its "Kähler form" $\omega_g$ satisfies the following second order differential equation:

$$\partial\bar{\partial}\omega_g^{n-1} = 0 \ .$$

The following result of Gauduchon (see [Ga]) shows that on any compact complex manifold there are a lot of Gauduchon metrics:

**Theorem 5.21** *Any hermitian metric on $X$ is conformally equivalent to a Gauduchon metric. If $n \geq 2$, then this Gauduchon metric is unique up to a positive factor.*

Let $g$ be a Gauduchon metric on $X$. By an idea of Hitchin, (see [Bh]), if $L$ is a holomorphic line bundle over $X$, the *degree of $L$ with respect to $\omega_g$* can be defined by

$$\deg_g(L) := \int_X \frac{i}{2\pi} F \wedge \omega_g^{n-1} \ ,$$

where $F$ is the curvature of any hermitian connection on $L$ compatible with $\bar{\partial}_L$. Since any two such forms differ by a $\partial\bar{\partial}$-exact form, $\deg_g(L)$ is independent of the choice of connection. If $d\omega_g = 0$, then $\deg_g(L)$ is the usual topological degree defined above, but in general, $\deg_g(L)$ is not a topological invariant. For a proof of the following result see [Bh]:

**Proposition 5.22** *Let $X$ be a compact complex surface. If $\omega$ is a positive $\partial\bar{\partial}$-closed (1,1)-form on $X$, then $\deg_\omega(L)$ depends only on the image of $c_1(L)$ in $H^2(X, \mathbb{R})$ if and only if $b_1(X)$ is even.*

*Remark.* By the results of Kodaira, Siu and Miyaoka, the condition $b_1(X)$ even is equivalent to the existence of a Kähler metric on the surface $X$.

Having defined the degree of holomorphic line bundles, the degree of a torsion-free coherent sheaf $\mathcal{E}$ is defined by

$$\deg_g(\mathcal{E}) := \deg_g(\det(\mathcal{E}))$$

(for $\det(\mathcal{E})$ see Definition 1.34), and the definition of $g$-stability (resp. $g$-semi-stability) can be repeated verbatim. Moreover, if $\mathcal{F}$ is a coherent torsion sheaf (i.e. $\mathrm{rk}(\mathcal{F}) = 0$), we still can define the degree of $\mathcal{F}$ by

$$\deg_g(\mathcal{F}) := \deg_g(\det(\mathcal{F})) \ ,$$

since $\det(\mathcal{F})$ is a line bundle.

**Lemma 5.23** *If $\mathcal{F}$ is a coherent torsion sheaf, then*

$$deg_g(\mathcal{F}) \geq 0 \ .$$

*Proof.* By Proposition 1.37, the line bundle $\det(\mathcal{F})$ has non-trivial holomorphic sections. If $s$ is a non-trivial section of $\det(\mathcal{F})$, then $Z_s$, the set of zeros of $s$, is either the empty set or a divisor of $X$. We get

$$\deg_g(\mathcal{F}) = \int_{Z_s} \omega_g^{n-1} = \mathrm{Vol}_g(Z_s) \geq 0$$

(see [G-H], p.171, [Ko3], p. 169, [Bh]).

We shall present, for the convenience of the reader, the elementary properties of stable (resp. semistable) sheaves following closely [O-S-S], Chapter II, 1 and [Ko3], Chapter V, 7, where more details can be found.

**Proposition 5.24** *Let $\mathcal{F}$ be a torsion-free coherent sheaf over a compact complex manifold $X$ with a Gauduchon metric $g$. Then:*

*(1) If $rk(\mathcal{F}) = 1$, then $\mathcal{F}$ is $g$-stable.*

*(2) Let $L$ be a line bundle over $X$. Then $\mathcal{F} \otimes L$ is $g$-stable (resp. $g$-semistable) if and only if $\mathcal{F}$ is $g$-stable (resp. $g$-semistable).*

*Proof.* Trivial.

**Proposition 5.25** *Let $\mathcal{F}$ be a torsion-free coherent sheaf over a compact complex manifold $X$ with a Gauduchon metric $g$. Then:*

*(1) $\mathcal{F}$ is $g$-semistable if and only if $\mu_g(\mathcal{F}) \leq \mu_g(\mathcal{F}'')$ for every quotient sheaf $\mathcal{F}''$ such that $0 < rk(\mathcal{F}'')$.*

*(2) $\mathcal{F}$ is $g$-stable if and only if $\mu_g(\mathcal{F}) < \mu_g(\mathcal{F}'')$ for every quotient sheaf $\mathcal{F}''$ such that $0 < rk(\mathcal{F}'') < rk(\mathcal{F})$.*

*Proof.* (cf. [O-S-S, Ko3]) If $\mathcal{F}''$ is a quotient sheaf of $\mathcal{F}$, then we denote by $\mathcal{F}'$ the kernel of the natural projection $\mathcal{F} \to \mathcal{F}''$ and we have an exact sequence

$$0 \to \mathcal{F}' \to \mathcal{F} \to \mathcal{F}'' \to 0 .$$

By Proposition 1.36, we get

$$\det(\mathcal{F}) \cong \det(\mathcal{F}') \otimes \det(\mathcal{F}'') ,$$

hence

$$\deg_g(\mathcal{F}) = \deg_g(\mathcal{F}') + \deg_g(\mathcal{F}'').$$

Let $r = rk(\mathcal{F})$, $r' = rk(\mathcal{F}')$ and $r'' = rk(\mathcal{F}'')$. Then $r = r' + r''$ and we can suppose $r' > 0$ (otherwise, $\mathcal{F}'$ is a torsion sheaf, hence it is zero since $\mathcal{F}' \subset \mathcal{F}$ is a torsion-free sheaf; then $\mathcal{F}'' \cong \mathcal{F}$). We obtain

$$r\mu_g(\mathcal{F}) = \deg_g(\mathcal{F}) = r'\mu_g(\mathcal{F}') + r''\mu_g(\mathcal{F}'') ,$$

or, equivalently

$$r'(\mu_g(\mathcal{F}) - \mu_g(\mathcal{F}')) + r''(\mu_g(\mathcal{F}) - \mu_g(\mathcal{F}'')) = 0 .$$

It follows $\mu_g(\mathcal{F}') \leq \mu_g(\mathcal{F})$ if and only if $\mu_g(\mathcal{F}) \leq \mu_g(\mathcal{F}'')$, hence (1), and $\mu_g(\mathcal{F}') < \mu_g(\mathcal{F})$ if and only if $\mu_g(\mathcal{F}) < \mu_g(\mathcal{F}'')$, hence (2).

**Proposition 5.26** *Let $\mathcal{F}$ be a torsion-free sheaf over a compact complex manifold $X$ with a Gauduchon metric $g$. Then:*

*(1) $\mathcal{F}$ is g-semistable if and only if either one of the following conditions holds:*

   *(a) $\mu_g(\mathcal{F}') \le \mu_g(\mathcal{F})$ for any subsheaf $\mathcal{F}'$ such that the quotient $\mathcal{F}/\mathcal{F}'$ is torsion-free;*

   *(b) $\mu_g(\mathcal{F}) \le \mu_g(\mathcal{F}'')$ for any torsion-free quotient sheaf $\mathcal{F}''$.*

*(2) $\mathcal{F}$ is g-stable if and only if either one of the following conditions holds:*

   *(c) $\mu_g(\mathcal{F}') < \mu_g(\mathcal{F})$ for any subsheaf $\mathcal{F}'$ such that the quotient $\mathcal{F}/\mathcal{F}'$ is torsion-free and $0 < rk(\mathcal{F}') < rk(\mathcal{F})$;*

   *(d) $\mu_g(\mathcal{F}) < \mu_g(\mathcal{F}'')$ for any torsion-free quotient sheaf $\mathcal{F}''$ such that $0 < rk(\mathcal{F}'') < rk(\mathcal{F})$.*

*Proof.* (cf. [O-S-S, Ko3]) For any exact sequence

$$0 \to \mathcal{F}' \to \mathcal{F} \to \mathcal{F}'' \to 0 ,$$

we take as in Section 1.4 $\mathcal{T}''$ to be the torsion subsheaf of $\mathcal{F}''$. Let $\mathcal{F}_1'' = \mathcal{F}''/\mathcal{T}''$ and take $\mathcal{F}_1'$ to be the kernel of the surjection $\mathcal{F} \to \mathcal{F}_1'' \to 0$. We get the exact sequence

$$0 \to \mathcal{F}_1' \to \mathcal{F} \to \mathcal{F}_1'' \to 0 ,$$

where all sheaves are torsion-free, and the exact sequence

$$0 \to \mathcal{F}' \to \mathcal{F}_1' \to \mathcal{T}'' \to 0 ,$$

where $\mathcal{T}''$ is a torsion sheaf. By Proposition 1.36 and Lemma 5.23, it follows

$$\deg_g(\mathcal{F}') \le \deg_g(\mathcal{F}_1') ,$$

hence

$$\mu_g(\mathcal{F}') \le \mu_g(\mathcal{F}_1')$$

since $rk(\mathcal{F}_1') = rk(\mathcal{F}') > 0$. Similarly, from the exact sequence

$$0 \to \mathcal{T}'' \to \mathcal{F}'' \to \mathcal{F}_1'' \to 0$$

we get

$$\mu_g(\mathcal{F}_1'') \le \mu_g(\mathcal{F}'') .$$

Now, from these inequalities, all the assertions follow in a standard way.

**Proposition 5.27** *Let $\mathcal{F}$ be a torsion-free sheaf over a compact complex manifold $X$ with a Gauduchon metric g. Then $\mathcal{F}$ is g-stable (resp. g-semistable) if and only if its dual $\mathcal{F}^*$ is g-stable (resp. g-semistable).*

*Proof.* (cf. [O-S-S, Ko3]) Since $\det(\mathcal{F}) = (\wedge^r \mathcal{F})^{**}$ by Corollary 1.31 it follows $\det(\mathcal{F})^* = (\wedge^r \mathcal{F})^*$. Since $\mathcal{F}$ is torsion-free the singularity set $S(\mathcal{F})$ is an analytic subset of $X$ of codimension at least 2 (Corollary 1.25). By definition $\det(\mathcal{F}^*) = (\wedge^r \mathcal{F}^*)^{**}$. Then we have a canonical isomorphism

$$f : (\textstyle\bigwedge^r \mathcal{F})^*|_{X \setminus S(\mathcal{F})} \longrightarrow (\textstyle\bigwedge^r \mathcal{F}^*)^{**}|_{X \setminus S(\mathcal{F})} \ .$$

Since $(\bigwedge^r \mathcal{F}^*)^{**}$ is reflexive (Corollary 1.31) and hence normal (Proposition 1.30), the sheaf

$$\mathcal{H}om((\textstyle\bigwedge^r \mathcal{F})^*, (\textstyle\bigwedge^r \mathcal{F}^*)^{**})$$

is also normal. Therefore, $f$ extends to a homomorphism

$$\tilde{f} : (\textstyle\bigwedge^r \mathcal{F})^* \longrightarrow (\textstyle\bigwedge^r \mathcal{F}^*)^{**} \ .$$

In the same way the inverse $h$ of $f$ extends to a homomorphism $\tilde{h}$ and $\tilde{f} \circ \tilde{h}$, $\tilde{h} \circ \tilde{f}$ follow easily identity homomorphisms. We get the isomorphism

$$(\det(\mathcal{F}))^* \cong \det(\mathcal{F}^*) \ ,$$

hence the equality

$$\mu_g(\mathcal{F}^*) = -\mu_g(\mathcal{F}) \ .$$

Now, assume that $\mathcal{F}^*$ is $g$-stable and consider an exact sequence

$$0 \to \mathcal{F}' \to \mathcal{F} \to \mathcal{F}'' \to 0$$

such that $\mathcal{F}''$ is torsion-free. Dualizing this sequence, we get the exact sequence

$$0 \to \mathcal{F}''^* \to \mathcal{F}^* \to \mathcal{F}'^* \ .$$

Then, by Proposition 5.26, we have

$$\mu_g(\mathcal{F}) = -\mu_g(\mathcal{F}^*) < -\mu_g(\mathcal{F}''^*) = \mu_g(\mathcal{F}'') \ .$$

Again, by Proposition 5.26, we get that $\mathcal{F}$ is $g$-stable.

Let $\mathcal{F}$ be $g$-stable and consider an exact sequence

$$0 \to \mathcal{F}' \to \mathcal{F}^* \to \mathcal{F}'' \to 0$$

such that $\mathcal{F}''$ is torsion-free. Dualizing it, we have the exact sequence

$$0 \to \mathcal{F}''^* \to \mathcal{F}^{**} \to \mathcal{F}'^* \ .$$

Consider $\mathcal{F}$ as a subsheaf of $\mathcal{F}^{**}$ under the natural injection $\mathcal{F} \to \mathcal{F}^{**}$ and define $\mathcal{F}_1$ by

$$\mathcal{F}_1 := \mathcal{F} \cap \mathcal{F}''^* \ .$$

Then we get an exact sequence

$$0 \to \mathcal{F}''^*/\mathcal{F}_1 \to \mathcal{F}^{**}/\mathcal{F} \to \mathcal{T}'' \to 0 \ .$$

Since $\mathcal{F}^{**}/\mathcal{F}$ is a torsion sheaf, so are $\mathcal{F}''^*/\mathcal{F}_1$ and $\mathcal{T}''$. By Proposition 1.37, $\det(\mathcal{F}^{**}/\mathcal{F})$ is a trivial line bundle, hence also $\det(\mathcal{F}''^*/\mathcal{F}_1)$ is a trivial line bundle. It follows

$$\det(\mathcal{F}''^*) = \det(\mathcal{F}_1) \ ,$$

hence

$$\mu_g(\mathcal{F}_1) = \mu_g(\mathcal{F}'''^*)$$

since $\mathcal{F}_1$ and $\mathcal{F}'''^*$ have the same rank. Then, by Proposition 5.26, we have

$$\mu_g(\mathcal{F}'') = -\mu_g(\mathcal{F}'''^*) = -\mu_g(\mathcal{F}_1) > -\mu_g(\mathcal{F}) = \mu_g(\mathcal{F}^*) \,,$$

hence $\mathcal{F}^*$ is $g$-stable. For the semistable case the proof is similar.

**Proposition 5.28** *Let $\mathcal{F}_1$ and $\mathcal{F}_2$ be $g$-semistable sheaves over a compact complex manifold $X$ with a Gauduchon metric $g$. Let $f : \mathcal{F}_1 \to \mathcal{F}_2$ be a homomorphism.*

*(1) If $\mu_g(\mathcal{F}_1) > \mu_g(\mathcal{F}_2)$, then $f = 0$;*

*(2) If $\mu_g(\mathcal{F}_1) = \mu_g(\mathcal{F}_2)$ and $\mathcal{F}_1$ is $g$-stable, then $rk(\mathcal{F}_1) = rk(f(\mathcal{F}_1))$ and $f$ is injective unless $f = 0$;*

*(3) If $\mu_g(\mathcal{F}_1) = \mu_g(\mathcal{F}_2)$ and if $\mathcal{F}_2$ is $g$-stable, then $rk(\mathcal{F}_2) = rk(f(\mathcal{F}_1))$ and $f$ is generically surjective unless $f = 0$.*

*Proof.* Assume $f \neq 0$ and denote $f(\mathcal{F}_1)$ by $\mathcal{F}$. Then $\mathcal{F}$ is a torsion-free quotient sheaf of $\mathcal{F}_1$.
(1) Since

$$\mu_g(\mathcal{F}) \leq \mu_g(\mathcal{F}_2) < \mu_g(\mathcal{F}_1) \leq \mu_g(\mathcal{F}) \,,$$

we get a contradiction. It follows $f = 0$.
(2) If $\mathcal{F}_1$ is $g$-stable and if $rk(\mathcal{F}_1) > rk(\mathcal{F})$, then

$$\mu_g(\mathcal{F}) \leq \mu_g(\mathcal{F}_2) = \mu_g(\mathcal{F}_1) < \mu_g(\mathcal{F}) \,,$$

which is imposible. Hence, $rk(\mathcal{F}_1) = rk(\mathcal{F})$ and $f$ is injective.
(3) If $\mathcal{F}_2$ is $g$-stable and if $rk(\mathcal{F}_2) > rk(\mathcal{F})$, then

$$\mu_g(\mathcal{F}) < \mu_g(\mathcal{F}_2) = \mu_g(\mathcal{F}_1) \leq \mu_g(\mathcal{F}) \,,$$

contradiction. Hence, $rk(\mathcal{F}_2) = rk(\mathcal{F})$ and $f$ is generically surjective.

**Corollary 5.29** *Let $X$ be a compact complex manifold with a Gauduchon metric $g$. Let $E_1$ and $E_2$ be $g$-semistable vector bundles over $X$, such that $rk(E_1) = rk(E_2)$ and $\deg_g(E_1) = \deg_g(E_2)$. If $E_1$ or $E_2$ is $g$-stable, then any non-zero sheaf homomorphism $f : E_1 \to E_2$ is an isomorphism.*

*Proof.* By Proposition 5.28, $f$ is an injective sheaf homomorphism and the induced homomorphism

$$\det(f) : \det(E_1) \longrightarrow \det(E_2)$$

is non-zero and injective (see Proposition 1.35). Now, consider $\det(f)$ as a holomorphic section of the line bundle

$$L = \mathcal{H}om(\det(E_1), \det(E_2)) \cong (\det(E_1))^{-1} \otimes (\det(E_2)) \,.$$

Since $\deg_g(L) = 0$, by Corollary 1 in [Bh], it follows that $\det(f)$ is an isomorphism ($L$ is trivial), hence $f$ is an isomorphism.

**Corollary 5.30** *If $\mathcal{F}$ is a g-semistable sheaf over a compact complex manifold $X$ with a Gauduchon metric $g$ such that $\deg_g(\mathcal{F}) < 0$, then $\mathcal{F}$ admits no non-zero holomorphic section.*

*Proof.* Let $\mathcal{O}_X$ be the sheaf of germs of holomorphic functions on $X$. Every holomorphic section of $\mathcal{F}$ is a sheaf homomorphism $f : \mathcal{O}_X \to \mathcal{F}$. Now, apply Proposition 5.28.

**Corollary 5.31** *Every g-stable vector bundle $E$ over a compact complex manifold $X$ with a Gauduchon metric $g$ is simple.*

*Proof.* Let $f : E \to E$ be an endomorphism of the vector bundle $E$ and let $a$ be an eigenvalue of $f(x) : E(x) \to E(x)$ at a point $x \in X$. Applying Corollary 5.29 to $f - a.id_E$, we get that $f = a.id_E$, i.e. $E$ is simple.

*Remark.* Let $X$ be a compact complex manifold with a Gauduchon metric $g$. Let $E$ be a $\mathcal{C}^\infty$-complex vector bundle over $X$ with $c_1(E) \in \mathrm{NS}(X)$. Let $\mathcal{M}_X^{s,E}$ be the moduli space of simple holomorphic structures in $E$ and let $\mathcal{M}_X^{g-st,E}$ be the set of $g$-stable holomorphic structures in $E$. By Corollaries 5.29 and 5.31, we get a natural injective map

$$I : \mathcal{M}_X^{g-st,E} \hookrightarrow \mathcal{M}_X^{s,E} \ .$$

In the next section we shall study this natural map.

For a proof of the following result, see [Ko3], p. 174 and [Bh]:

**Lemma 5.32** *Let $X$ be a compact complex manifold with a Gauduchon metric $g$. Then, for a given torsion-free coherent sheaf $\mathcal{F}$ over $X$, there is an integer $m_0$ such that*

$$\mu_g(\mathcal{F}') \leq m_0$$

*for all coherent subsheaves $\mathcal{F}'$ of $\mathcal{F}$.*

**Lemma 5.33** *Let $X$ be a compact complex manifold with a Gauduchon metric $g$. For a given torsion-free coherent sheaf $\mathcal{F}$, there is a unique subsheaf $\mathcal{F}_1$ with torsion-free quotient $\mathcal{F}/\mathcal{F}_1$ such that, for every subsheaf $\mathcal{F}'$ of $\mathcal{F}$*

*(1) $\mu_g(\mathcal{F}') \leq \mu_g(\mathcal{F}_1)$ ;*

*(2) $rk(\mathcal{F}') \leq rk(\mathcal{F}_1)$ if $\mu_g(\mathcal{F}') = \mu_g(\mathcal{F}_1)$ .*

*Moreover, $\mathcal{F}_1$ is g-semistable.*

*Proof.* (cf. [Ko3], p. 175). The existence of $\mathcal{F}_1$ follows from Lemma 5.32 From (1) we get that $\mathcal{F}_1$ is $g$-semistable. Now we prove the uniqueness. Let $\mathcal{F}_1'$ be another subsheaf satisfying (1) and (2) and let $p : \mathcal{F} \to \mathcal{F}/\mathcal{F}_1'$ be the projection. If $p(\mathcal{F}_1) = 0$, then $\mathcal{F}_1 \subset \mathcal{F}_1'$ and from (2) we get $\mathcal{F}_1 = \mathcal{F}_1'$. Let us suppose $p(\mathcal{F}_1) \neq 0$ and define a subsheaf $\mathcal{F}'$ of $\mathcal{F}_1$ by the exact sequence

$$0 \to \mathcal{F}' \to \mathcal{F}_1 \to p(\mathcal{F}_1) \to 0 \ .$$

Since $\mathcal{F}_1$ is $g$-semistable, we get

$$\mu_g(\mathcal{F}') \leq \mu_g(\mathcal{F}_1) \leq \mu_g(p(\mathcal{F}_1)) \ .$$

On the other hand, we have an exact sequence

$$0 \to \mathcal{F}_1' \to p^{-1}(p(\mathcal{F}_1)) \to p(\mathcal{F}_1) \to 0 \ .$$

Since $p(\mathcal{F}_1)$ is non-zero and torsion-free, by (1) and (2) for $\mathcal{F}_1'$, we get

$$\mu_g(p^{-1}(p(\mathcal{F}_1))) < \mu_g(\mathcal{F}_1') \ .$$

This implies (by computation)

$$\mu_g(p(\mathcal{F}_1)) < \mu_g(\mathcal{F}_1') = \mu_g(\mathcal{F}_1) \ ,$$

contradiction. It follows $\mathcal{F}_1 = \mathcal{F}_1'$.

By repeated application of the Lemma 5.33, we get the Harder-Narasimhan filtration (see [H-N, Mr3], [Ko3], p.174):

**Theorem 5.34** *Given a torsion-free coherent sheaf $\mathcal{F}$ over a compact complex manifold $X$ with a Gauduchon metric $g$, there is a unique filtration by subsheaves*

$$0 = \mathcal{F}_0 \subset \mathcal{F}_1 \subset \cdots \subset \mathcal{F}_{s-1} \subset \mathcal{F}_s = \mathcal{F}$$

*such that, for $1 \leq i \leq s-1$, the sheaf $\mathcal{F}_i/\mathcal{F}_{i-1}$ is the maximal $g$-semistable subsheaf of $\mathcal{F}/\mathcal{F}_{i-1}$.*

Now, we shall present some examples of $g$-stable vector bundles.

(1) Let $X$ be a compact complex manifold and let $g$ be any Gauduchon metric on $X$. Let $E$ be an irreducible holomorphic vector bundle over $X$. Then, just by the definition, $E$ is $g$-stable. In fact, we can say that $E$ is *absolutely stable*, i.e. $g$-stable for any Gauduchon metric $g$ on $X$. Since we gave examples of irreducible holomorphic vector bundles over nonalgebraic surfaces, we have examples of $g$-stable vector bundles in this case.

(2) Let $X$ be a non-kählerian compact complex surface. We shall present $g$-stable holomorphic 2-vector bundles constructed by V. Vuletescu (see [Vu1]). Let $g$ be a Gauduchon metric on $X$. One knows that the map

$$\deg_g : \mathrm{Pic}_0(X) \longrightarrow \mathbb{R}$$

is a surjective homomorphism of real Lie groups and, if a holomorphic vector bundle $E$ has a non-zero section, then $\deg_g(E) \geq 0$. Moreover, if $L$ is a holomorphic line bundle such that $H^0(X, L) \neq 0$, then $\deg_g(L) \geq 0$ with equality if and only if $L$ is trivial (see [Bh]).

**Lemma 5.35** *Let $X$ be a non-kählerian compact complex surface and let $g$ be a Gauduchon metric on $X$. Then the set*

$$\mathcal{D} := \{\deg_g(\mathcal{O}_X(D)) \mid D \text{ an irreducible curve on } X \}$$

*is finite.*

*Proof.* (cf. [Vu1]) If the algebraic dimension of $X$ is zero then, by Theorem 2.16, the number of irreducible curves on $X$ is finite, so $\mathcal{D}$ is finite. If the algebraic dimension of $X$ is one then, by Theorem 2.13 and Proposition 2.14, there is a connected morphism $\pi : X \to B$ onto a nonsingular projective curve $B$ and any irreducible curve on $X$ is contained in a fibre of $\pi$. Thus, we have

$$\{D \mid D \text{ irreducible curve on } X\} =$$

$$= \{D \mid D \text{ is contained in a singular fibre of } \pi\}\cup$$

$$\cup\{D \mid D \text{ is a smooth fibre of } \pi\} \ .$$

Since the first member of this union is finite, it will be sufficient to study the second member of the union. Let $D_1$, $D_2$ be two smooth fibres of $\pi$ and let $b_i = \pi(D_i)$, $i = 1, 2$. Denote by $\mathcal{L} := \mathcal{O}_B(b_1 - b_2)$ and fix a line bundle $\mathcal{H}$ over $B$ such that $\deg(\mathcal{H}) \geq 2g' - 1$, where $g'$ is the genus of the curve $B$. Then, for any integer $n$, we have by Riemann-Roch formula on $B$

$$H^0(B, \mathcal{H} \otimes \mathcal{L}^{\otimes n}) \neq 0 \ ,$$

i.e.

$$H^0(X, \pi^*(\mathcal{H}) \otimes (\mathcal{O}_X(D_1 - D_2))^{\otimes n}) \neq 0 \ ,$$

hence (by the proof of Lemma 5.23)

$$n(\deg_g(\mathcal{O}_X(D_1)) - \deg_g(\mathcal{O}_X(D_2))) + \deg_g(\pi^*(\mathcal{H})) \geq 0 \ ,$$

for any $n \in \mathbb{Z}$. It follows that

$$\deg_g(\mathcal{O}_X(D_1)) = \deg_g(\mathcal{O}_X(D_2)) \ ,$$

hence $\mathcal{D}$ is finite.

Let

$$\varepsilon_g(X) := \min\{\deg_g(\mathcal{O}_X(D)) \mid D \text{ irreducible curve on } X\}$$

if there exists at least one irreducible curve on $X$ and $\varepsilon_g(X) = \infty$, otherwise. Let $\tilde{D}$ be the abelian subgroup of $\mathbb{R}$ generated by the finite set $\mathcal{D}$ of Lemma 5.35.

**Lemma 5.36** *Let $X$ be a non-kählerian compact complex surface and let $g$ be a Gauduchon metric on $X$. Then there exists a holomorphic line bundle $L \in Pic_0(X)$, with the following properties:*

*(1)* $\deg_g(L) \in (0, \varepsilon_g(X))$;

*(2)* $\deg_g(L) \notin \tilde{D}$;

*(3)* $H^0(X, L^{\otimes 2}) = H^2(X, L^{\otimes 2}) = H^2(X, (L^*)^{\otimes 2}) = 0$ .

*Proof.* (cf. [Vu1]) If the algebraic dimension of $X$ is zero, then the condition

$$H^0(X, L^{\otimes 2}) \neq 0 \; , \; \text{or} \; H^2(X, L^{\otimes 2}) \neq 0 \; , \; \text{or} \; H^2(X, (L^*)^{\otimes 2}) \neq 0$$

is satisfied for an at most countable subset of $\text{Pic}_0(X)$. Since $\tilde{D}$ is at most countable and since $\deg_g^{-1}(0, \varepsilon_g(X))$ is open, it follows that there exists a holomorphic line bundle $L \in \text{Pic}_0(X)$ with the above properties (1)-(3).

If the algebraic dimension of $X$ is one, then one knows that the canonical line bundle $\mathcal{K}_X$ has the form

$$\mathcal{K}_X \cong \mathcal{O}_X(\textstyle\sum m_i D_i) \; ,$$

where $D_i$ are irreducible curves on $X$ and $m_i \in \mathbb{Z}$ (see [B-P-V], p. 161). It follows that any holomorphic line bundle $L \in \text{Pic}_0(X)$ with

$$\deg_g(L) \notin \tilde{D} \text{ and } 0 < \deg_g(L^{\otimes 2}) < \varepsilon_g(X)$$

satisfies the properties (1)-(3).

**Theorem 5.37** *Let $X$ be a non-kählerian compact complex surface and $g$ a Gauduchon metric on $X$. Let $c$ be an arbitrary positive integer. Then there exists a holomorphic 2-vector bundle $E$ over $X$, which is $g$-stable, with trivial determinant and the second Chern class equal to $c$.*

*Proof.* (cf. [Vu1]) Let $L \in \text{Pic}_0(X)$ as in Lemma 5.36 and let $Y \subset X$ be a locally complete intersection of codimension 2 in $X$ with length$(Y) = c > 0$. By Serre construction (Theorem 4.3), there exists a holomorphic rank two vector bundle $E$ given by an extension

$$0 \to L^* \to E \to J_Y \otimes L \to 0 \; .$$

We have

$$\det(E) \cong L^* \otimes L \cong \mathcal{O}_X \; , \; c_2(E) = \ell(Y) = c > 0 \; .$$

By Theorem 4.34, it follows that $E$ is simple. To prove that $E$ is $g$-stable, by Proposition 5.26, it suffices to show that for any holomorphic line bundle $L' \hookrightarrow E$ we have $\deg_g(L') < 0$. From $L' \hookrightarrow E$ we get $L' \hookrightarrow L$ or $L' \hookrightarrow L^*$. In the second case, we get

$$\deg_g((L')^* \otimes L^*) \geq 0 \; ,$$

i.e.

$$\deg_g(L') \leq \deg_g(L^*) < 0 \; .$$

In the first case, we have $L = L'$ or $L = L' \otimes \mathcal{O}_X(\sum m_i D_i)$, with $D_i$ irreducible curves on $X$ and $m_i > 0$. If $L = L'$, then we get a non-zero non-constant endomorphism $E \to L \to E$ (since $\det(E) \cong \mathcal{O}_X$, $E^* \cong E$), therefore $E$ is not simple, contradiction. If $L = L' \otimes \mathcal{O}_X(\sum m_i D_i)$, then we obtain, as in the second case, that $\deg_g(L') < 0$.

*Remark.* See, also [Bm-Hu, Ml, Pl, Vu3, Vu4], for other results on vector bundles over nonalgebraic surfaces.

## 5.4 Moduli spaces of stable vector bundles

Let $X$ be a compact complex manifold endowed with a Gauduchon metric $g$. Let $E$ be a $\mathcal{C}^\infty$-complex vector bundle over $X$ with $c_1(E) \in \mathrm{NS}(X)$. Consider the set $\mathcal{M}_X^{g-st,E}$ of isomorphism classes of holomorphic $g$-stable structures in $E$ and the functor

$$M_X^{g-st,E} : An \longrightarrow Sets \,,$$

which associates to a complex space $S$ the set of equivalence classes of families of holomorphic vector bundles $\mathcal{E}$ over $S \times X$, such that

(1) the restriction $\mathcal{E}_s$ of $\mathcal{E}$ to the fibre $X_s = \{s\} \times X$ for every $s \in S$ is topologically (smoothly) isomorphic to $E$;

(2) $\mathcal{E}_s$ is $g$-stable for every point $s \in S$.

For the following fundamental result, see [Mr3, Gi1]:

**Theorem 5.38** *Let $X$ be an algebraic surface and let $H$ be a very ample divisor on $X$. Then there exists a quasi-projective variety, whose underlying complex space $\mathcal{M}_X^{H-st,E}$ is a coarse moduli space for the functor $M_X^{H-st,E}$.*

Very briefly, the proof runs as follows: the $H$-stable vector bundles $F$ with fixed Chern classes

$$c_1(F) = c_1(E) = c_1 \text{ and } c_2(F) = c_2(E) = c_2$$

form a *bounded family* of coherent sheaves. Then, there exists an integer $n_0$, depending only on $X, H, c_1$ and $c_2$, such that for all $n \geq n_0$, the bundle $F \otimes H^{\otimes n}$ is generated by its global sections and has no higher cohomology. Thus, for an appropriate $m$, which does not depend on $F$, we get that $F$ is a quotient of

$$G := \left( H^{\otimes(-n)} \right)^{\oplus m} \,.$$

In this way all of the above $H$-stable vector bundles $F$ are quotients of the fixed bundle $G$ and these quotients can be parametrized by the Grothendieck Quot scheme (see, for example, [Kl]). We may restrict to an open set $Q$ in the Grothendieck Quot scheme. The group

$$GL(m, \mathbb{C}) = \mathrm{Aut}(G)$$

acts on $Q$ in a natural way, and we can see that two stable vector bundles are isomorphic if and only if they are in the same orbit of $GL(m, \mathbb{C})$. The more difficult step of the proof involves using geometric invariant theory to construct an appropriate quotient of $Q$ by $GL(m, \mathbb{C})$. Finally, one shows that the scheme so constructed coarsely represents the moduli functor. A modification of the arguments shows that the complex space associated to this scheme coarsely represents the moduli functor $M_X^{H-st,E}$ from complex spaces to sets. By the uniqueness property of coarse moduli spaces, the complex space associated to this scheme is $\mathcal{M}_X^{H-st,E}$

Gieseker [Gi1] and Maruyama [Mr3] constructed a "compactification" of this moduli space, i.e. a projective variety, parametrising certain equivalence classes of semi-stable sheaves.

By a result of Wehler [Wh], the germ of $\mathcal{M}_X^{H-st,E}$ at any point $[F]$ is the base of the versal deformation of $F$. Locally, there always exist on $\mathcal{M}_X^{H-st,E}$ universal bundles but, in general, these bundles do not fit together to a global universal bundle over $\mathcal{M}_X^{H-st,E} \times X$ (i.e. $\mathcal{M}_X^{H-st,E}$ is not generally a fine moduli space).

We know, by general deformation theory (see Theorem 5.4), that the Zariski tangent space of $\mathcal{M}_X^{H-st,E}$ at a point $[F]$ is $H^1(X, \mathcal{E}nd(F))$. From the decomposition (see Section 5.1)

$$\mathcal{E}nd(F) \cong \mathcal{O}_X \oplus sl(F)$$

we get

$$H^1(X, \mathcal{E}nd(F)) = H^1(X, \mathcal{O}_X) \oplus H^1(X, sl(F)) \ .$$

The space $H^1(X, sl(F))$ is the Zariski tangent space of the base of the versal deformation of the projective bundle $\mathbb{P}(F)$ associated to $F$. In fact, $\mathcal{M}_X^{H-st,E}$ is locally a product of the base of this versal deformation with $\mathrm{Pic}_0(X)$ (see [E-F] and Theorem 5.6). The base of the versal deformation of $\mathbb{P}(F)$ can be described as the inverse image of $0 \in H^2(X, sl(F))$ under the Kuranishi map

$$K_{[F]} : H^1(X, sl(F)) \longrightarrow H^2(X, sl(F)) \ .$$

This Kuranishi description of the germ $(\mathcal{M}_X^{H-st,E}, [F])$, together with Riemann-Roch formula, give the following bound for the dimension of $\mathcal{M}_X^{H-st,E}$ at $[F]$:

$$\dim_{[F]} \left( \mathcal{M}_X^{H-st,E} \right) \geq 4c_2(E) - c_1^2(E) - 3\chi(\mathcal{O}_X) + q(X) \ .$$

The bound

$$4c_2(E) - c_1^2(E) - 3\chi(\mathcal{O}_X) + q(X) = 8\Delta(E) - 3\chi(\mathcal{O}_X) + q(X)$$

is called the "expected dimension" of $\mathcal{M}_X^{H-st,E}$.

The moduli space $\mathcal{M}_X^{H-st,E}$ is contained as open subset in the moduli space $\mathcal{M}_X^{s,E}$ of simple holomorphic structures. By Proposition 5.28 and Corollary 5.29, it follows that $\mathcal{M}_X^{H-st,E}$ is globally Hausdorff (see the criterion of separability from [L-O1]). The existence of stable vector bundles in the algebraic context and the structure of the moduli spaces $\mathcal{M}_X^{H-st,E}$ have been studied by many authors; a short list of references is the following: [An, Bt, Bu1, Bu2, Bo2, Bs, Cd, Do2, Do3, D-K, Dz1, Dz2, Dz3, D-L, Fr, F-M3, Gi1, Gi3, G-L, Gt, Gt-H, Gd1, Gd2, H-N, Hh2, H-L, H-Sp, Hu1, Hu2, Hy, LP1, Li1, Li2, L-O2, Mr1, Mr2, Mr3, Mr4, Mr5, Mr6, Mr7, Mr8, Mu1, Mu2, Na, N-S, Nw, O-V1, O-V3, O-V4, Q1, Q2, Q3, Q4, Q5, Sn, Sd, St1, St2, Tk, Tb1, Tb2, Tt, Um1, Um2, Zu].

Let $X$ be a compact complex manifold and let $g$ be a hermitian metric on $X$. Let $\omega_g$ be its "Kähler form" and let us denote as usually by $\Lambda_g$ the $L^2$-adjoint (the contraction with the metric) of the multiplication operator $x \mapsto x \wedge \omega_g$. If $(E, h)$ is a hermitian $r$-vector bundle over the hermitian manifold $(X, g)$ and, if $A$ is a unitary connection in $E$, then $F_A$ denotes the curvature form of the connection $A$.

**Definition 5.39** Let $(X, g)$ be a hermitian compact complex manifold and $(E, h)$ a hermitian $r$-vector bundle on it. A unitary connection $A$ in $E$ is called *weakly Hermitian-Einstein (w.H-E)* if $F_A$ has type (1,1) and if it exists a real function $\varphi_A$ on $X$ such that

$$\Lambda_g F_A = i\varphi_A.Id_E \ .$$

If $\varphi_A = C_A$ is a constant function, then $A$ is called *Hermitian-Einstein (H-E)* and this constant is called the *Einstein constant of A*.

This concept has been introduced by Kobayashi ( [Ko1]) as a generalisation of the notion of a Kähler-Einstein metric. We mention that if a connection $A$ is w.H-E then, by definition, its second component $\bar{\partial}_A$ is integrable (see [L-O1]), so it defines a holomorphic structure $E_A$ in $E$. If $g$ is a Gauduchon metric and if $A$ is H-E then, the Einstein constant $C_A$ is proportional to the slope of $(E, \bar{\partial}_A)$:

$$C_A = \frac{2\pi}{(n-1)!Vol_g(X)}\mu_g(E, \bar{\partial}_A) \ ,$$

where $n$ is the dimension of $X$ (see [Ko3], p. 178).

For the following result see [Bo2, Lu1], [Ko3], p. 114:

**Theorem 5.40 (Lübke)** *Let $(X, g)$ be a compact complex hermitian manifold of dimension $n$ with "Kähler form" $\omega_g$. Let $(E, h)$ be an hermitian vector bundle of rank $r$ over $X$. If $(E, h)$ is w.H-E, then*

$$\int_X ((r-1)c_1^2(E) - 2rc_2(E)) \wedge \omega_g^{n-2} \leq 0 \ ,$$

*and the equality holds if and only if $(E, h)$ is projectively flat (i.e. $\mathbb{P}(E)$ admits a flat $PU(r)$-structure).*

If $X$ is a compact complex surface, we obtain the inequality:

$$\Delta(E) \geq 0 \ ,$$

for any w.H-E vector bundle.

The following vanishing theorem is due to Kobayashi [Ko3], p. 52:

**Theorem 5.41 (Kobayashi)** *If a holomorphic vector bundle $E$ over a compact complex hermitian manifold $X$ possesses a compatible H-E connection $A$ with non-positive Einstein constant, then any section $s \in H^0(X, E)$ is A-parallel (A-covariantly constant). If $C_A$ is negative, then $H^0(X, E) = 0$.*

A H-E connection is called *reducible* if its holonomy is contained in a subgroup of type $U(k) \times U(r-k)$ for some $0 < k < r$ $(r = \text{rk}(E))$. This means that the vector bundle splits as an orthogonal sum of H-E vector bundles with the same constant. Irreducible H-E connections $A$ define simple holomorphic vector bundles $E_A$ since, by Kobayashi vanishing Theorem, every holomorphic section of $\mathcal{E}nd(E_A)$ must be parallel.

The stability of H-E vector bundles was proved by Kobayashi [Ko2] and Lübke [Lu2]:

**Theorem 5.42** *Let $(E, h)$ be an H-E vector bundle over a compact complex manifold $X$ endowed with a Gauduchon metric $g$. Then $E$ is $g$-semistable and $(E, h)$ is a direct sum*

$$(E, h) = (E_1, h_1) \oplus \cdots \oplus (E_k, h_k)$$

of g-stable H-E vector bundles $(E_1, h_1), \cdots, (E_k, h_k)$ with the same constant $C$ as $(E, h)$.

The converse has been conjectured independently by Kobayashi and Hitchin. The Kobayashi-Hitchin correspondence relates the complex geometry concept of "stable holomorphic vector bundle" to the differential geometry concept of "H-E connection":

**Theorem 5.43** Let $(X, g)$ be a compact complex manifold endowed with a Gauduchon metric. A holomorphic vector bundle $E$ over $X$ is g-stable if and only if it allows a hermitian metric $h$ such that the associated Chern connection $A_h$ is H-E and irreducible. This metric is unique up to a constant factor.

The existence of H-E connections in stable vector bundles was proved by Donaldson [Do2] for projective surfaces and later for projective manifolds, by Uhlenbeck and Yau [U-Y] for Kähler manifolds, by Buchdahl [Bh] for Gauduchon surfaces and finally, by Li and Yau [L-Y] for general Gauduchon manifolds.

Let $X$ be a compact complex manifold endowed with an arbitrary hermitian metric, fix a $\mathcal{C}^\infty$ hermitian vector bundle $(E, h)$ over $X$ and let $\mathcal{M}_X^{H-E,(E,h)}$ be the set of equivalence classes of irreducible H-E connections in $(E, h)$, modulo the action of the gauge group of unitary automorphisms $U(E, h)$.

The above theorem can be formulated as follows:

**Corollary 5.44** Let $(X, g)$ be an arbitrary compact hermitian manifold. Then the natural map $A \mapsto \bar{\partial}_A$ induces an injection

$$\tilde{I}_g : \mathcal{M}_X^{H-E,(E,h)} \longrightarrow \mathcal{M}_X^{s,E} .$$

If $g$ is Gauduchon, then the image of this map is precisely the subset $\mathcal{M}_X^{g-st,E}$ of $\mathcal{M}_X^{s,E}$, consisting of g-stable vector bundles.

Let us present the following "amusing proof" of the Bănică-Le Potier inequality (Theorem 4.17); see [Br4]:

**Corollary 5.45** Let $X$ be a nonalgebraic surface and let $E$ be a holomorphic vector bundle over $X$. Then $\Delta(E) \geq 0$.

*Proof.* There exists (obviously) a filtration of $E$ with coherent subsheaves

$$0 \subset \mathcal{F}_1 \subset \cdots \subset \mathcal{F}_k = E , \; k \leq r = \mathrm{rk}(E) ,$$

such that $G_i = \mathcal{F}_i / \mathcal{F}_{i-1}$ is a torsion-free and irreducible sheaf of rank $r_i$, $0 < r_i \leq r$. Then

$$\mu(E) = \frac{1}{r} \sum r_i \mu_i , \; \mu_i = c_1(G_i)/r_i ,$$

and

$$\chi(E) = \sum \chi(G_i) .$$

By Riemann-Roch formula we have

$$\Delta(E) = P(\mu) - \frac{1}{r}\sum r_i P(\mu_i) + \frac{1}{r}\sum r_i \Delta(G_i)$$

(see the notations of Section 4.2 ). But $G_i^{**}$ is an irreducible vector bundle and $\Delta(G_i^{**}) \leq \Delta(G_i)$. Since $G_i^{**}$ is irreducible it follows that $G_i^{**}$ is $g$-stable with respect to any Gauduchon metric $g$. By the Theorem 5.43, we have that $G_i^{**}$ admits a H-E connection. Then, by the Theorem 5.40, it follows that $\Delta(G_i^{**}) \geq 0$. Now, from the "concavity" property of the function $P$ (Section 4.2), we obtain $\Delta(E) \geq 0$.

For the following result see [Te1]:

**Theorem 5.46** *Let $(X, g)$ be a compact complex hermitian $n$-dimensional manifold and $(E, h)$ a fixed $C^\infty$ hermitian $r$-vector bundle over $X$. Then the map*

$$\tilde{I}_g : \mathcal{M}_X^{H-E,(E,h)} \longrightarrow \mathcal{M}_X^{s,E}$$

*induced by the natural projection $A \mapsto \bar{\partial}_A$ is the set theoretical support of a real analytic open embedding.*

The problem was studied in the Kähler case by Fujiki and Schumacher [F-S] and Miyajima [Mj].

For an algebraic surface $X$, the Theorem 5.38 gives the existence of a coarse moduli space $\mathcal{M}_X^{H-st,E}$ for stable vector bundles. For nonalgebraic surfaces $X$, the Theorem 5.46 together with the set-theoretical Kobayashi-Hitchin correspondence (Theorem 5.43), will give immediately the existence of a coarse moduli space $\mathcal{M}_X^{g-st,E}$ for $g$-stable vector bundles (see [Te1]):

**Corollary 5.47** *For any $C^\infty$ vector bundle $E$ over a compact complex manifold $X$ endowed with a Gauduchon metric $g$, it exists a coarse moduli space $\mathcal{M}_X^{g-st,E}$ for $g$-stable vector bundles with $C^\infty$ support $E$. This moduli space is an open Hausdorff subspace of $\mathcal{M}_X^{s,E}$.*

*Proof.* (cf. [Te1]) Define $\mathcal{M}_X^{g-st,E}$ to be the image of the open embedding given by Theorem 5.46 and construct the corresponding functor

$$M_X^{g-st,E}(\cdot) \to \mathrm{Hom}(\cdot, \mathcal{M}_X^{g-st,E}) ,$$

by restricting the corresponding classifying functor for simple vector bundles. The obtained functor is versal, because for any $[F] \in \mathcal{M}_X^{s,E}$, the germ of $\mathcal{M}_X^{s,E}$ in $[F]$ is the base of a versal deformation of $F$. On the other hand, if $F_i$, $i = 1, 2$, are $g$-stable non-isomorphic vector bundles such that $\deg_g(F_1) \leq \deg_g(F_2)$, then any morphism $F_2 \to F_1$ vanishes (Proposition 5.28 and Corollary 5.29). Therefore, the isomorphism classes $[F_1]$ and $[F_2]$ can be separated (see [L-O1]).

*Remark.* For applications of moduli spaces of stable vector bundles to the theory of 4-dimensional real manifolds see [D-K, F-M3, O-V4].

## 5.5 Vector bundles over ruled surfaces

Stable 2-vector bundles over ruled surfaces were studied by many authors; see, for example [Tk, H-Sp, Q3] . In this section we shall study algebraic 2-vector bundles over ruled surfaces, but we adopt another point of view: we shall construct moduli spaces of (algebraic) 2-vector bundles over a ruled surface $X$, which are defined independent of any ample divisor (line bundle) on $X$, by taking into account the special geometry of a ruled surface (see [B-S1, B-S2, Br1] and also [Bs, Wa]).

Let $C$ be a nonsingular curve of genus $g$ over the complex numbers field $\mathbb{C}$ and let $\pi : X \to C$ be a ruled surface over $C$. We have $X \cong \mathbb{P}(\mathcal{E})$, where $\mathcal{E}$ is a normalized locally free sheaf of rank 2 on $C$ (see [Hh1], p. 370). Let us denote by $\mathbf{e}$ the divisor on $C$ corresponding to the invertible sheaf $\bigwedge^2 \mathcal{E}$ and by $e = -\deg(\mathbf{e})$. Fix a section $C_0$ of $\pi$ with $\mathcal{O}_X(C_0) \cong \mathcal{O}_{\mathbb{P}(\mathcal{E})}(1)$ and $p_0$ a point of $C$. Let $f_0$ be the fibre $\pi^{-1}(p_0)$. Any element of the Neron-Severi group $\mathrm{NS}(X) \cong H^2(X, \mathbb{Z})$ can be written in the form $aC_0 + bF_0$, with $a, b \in \mathbb{Z}$ and $C_0^2 = -e$, $C_0 f_0 = 1$, $f_0^2 = 0$. Since the canonical divisor

$$K_X \sim -2C_0 + \pi^*(K_C + \mathbf{e}) ,$$

it follows that for numerical equivalence we have

$$K_X \equiv -2C_0 + (2g - 2 - e)f_0$$

(cf. [Hh1], p. 373). We will denote by $\mathcal{O}_C(1)$ the invertible sheaf associated to the divisor $p_0$ on $C$. If $L$ is an element of $\mathrm{Pic}(C)$, we shall write $L = \mathcal{O}_C(k) \otimes L_0$, where $k = \deg(L)$ and $L_0 \in \mathrm{Pic}_0(C)$. We also denote by

$$\mathcal{F}(aC_0 + bf_0) := \mathcal{F} \otimes \mathcal{O}_X(a) \otimes \pi^*(\mathcal{O}_C(b))$$

for any sheaf $\mathcal{F}$ over $X$ and any $a, b \in \mathbb{Z}$, where $\mathcal{O}_X(a) \cong \mathcal{O}_X(aC_0)$.

Let $E$ be an algebraic 2-vector bundle over $X$ with fixed numerical (topological) Chern classes $c_1 = (\alpha, \beta)$, $c_2 = \gamma$, where $\alpha, \beta, \gamma \in \mathbb{Z}$ and $c_1(E) \equiv \alpha C_0 + \beta f_0$, $c_2 = \deg(c_2(E)) = \gamma (\equiv$ is the numerical equivalence). Since the fibres of $\pi$, are isomorphic to $\mathbb{P}^1$, we can speak about the generic splitting type of $E$ and we have

$$E|_f \cong \mathcal{O}_f(d) \oplus \mathcal{O}_f(d')$$

for a "general" fibre $f$ of $\pi$, where $d \geq d'$ and $d + d' = \alpha$ (see [Gk1]). The integer $d$ is the first numerical invariant in our considerations.

The second numerical invariant $r$ is obtained by the following normalization:

$$-r = \inf\{\ell \in \mathbb{Z} \mid \exists L \in \mathrm{Pic}(C), \deg(L) = \ell \text{ s.t. } H^0(X, E(-dC_0) \otimes \pi^*(L)) \neq 0\} .$$

One has

$$H^0(X, E(-dC_0) \otimes \pi^*(L)) \cong H^0(C, \pi_*(E(-dC_0)) \otimes L)$$

and, moreover, the last cohomology group does not vanish when $\deg(L) \gg 0$ and it is zero when $\deg(L) \ll 0$ (for, use a suitable filtration of $\pi_*(E(-dC_0))$ with subbundles and Riemann-Roch for divisors on $C$). Therefore, there exists such an integer $r$.

Let us denote by $M(\alpha, \beta, \gamma, d, r) = M$ the set of isomorphism classes of algebraic 2-vector bundles over $X$, with fixed numerical Chern classes $c_1 = (\alpha, \beta), c_2 = \gamma$ and fixed type $(d, r)$. In particular, let us denote $M(0, 0, 0, d, r)$ by $M(d, r)$ (see [B-S1]). We have:

**Lemma 5.48** *For every $E \in M(\alpha, \beta, \gamma, d, r) = M$ there exist $L_1, L_2 \in Pic_0(C)$ and $Y \subset X$, a locally complete intersection of codimension 2 in $X$ or the empty set, such that the bundle $E$ is given by an extension*

$$0 \to \mathcal{O}_X(dC_0 + rf_0) \otimes \pi^*(L_2) \to E \to J_Y \otimes \mathcal{O}_X(d'C_0 + sf_0) \otimes \pi^*(L_1) \to 0, \quad (5.1)$$

*where $d + d' = \alpha$, $d \geq d'$, $r + s = \beta$ and*

$$\deg(Y) = \gamma + \alpha(de - r) - \beta d + 2dr - d^2 e .$$

*Proof.* For every $E \in M$ there exists $L_2^{-1} \in Pic_0(C)$ such that $H^0(X, E(-dC_0 - rf_0) \otimes \pi^*(L_2^{-1})) \neq 0$ and let $s$ be a non-zero section. The zero divisor associated with $s$ is $Z \sim mC_0 + \pi^{-1}(A)$, where $A$ is a divisor on $C$ of degree $n$ and $m \geq 0$. It follows that

$$\mathcal{O}_X(Z) \cong \mathcal{O}_X(mC_0 + nf_0) \otimes \pi^*(N_0)$$

with $N_0 \in Pic_0(C)$. By a classical argument (see Proposition 4.1), the section $s$ defines a section $s'$ in $H^0(X, E_1)$, where

$$E_1 = E((-d - m)C_0 + (-r - n)f_0) \otimes \pi^*(L_2^{-1} \otimes N_0^{-1}) ,$$

such that $s'$ vanishes only in a finite set of points $Y$ (a locally complete intersection of codimension 2 in $X$ or the empty set). The Koszul complex will give the resolution

$$0 \to \bigwedge^2 (E_1^*) \to E_1^* \xrightarrow{s^*} \mathcal{O}_X \to \mathcal{O}_Y \to 0 .$$

Tensoring by $\det(E_1)$, since $E_1 \cong E_1^* \otimes \det(E_1)$, we obtain

$$0 \to \mathcal{O}_X \to E_1 \to J_Y \otimes \det(E_1) \to 0 ,$$

where $J_Y$ is the ideal sheaf of $Y$ in $\mathcal{O}_X$. Finally, we get for the bundle $E$ the exact sequence

$$0 \to \mathcal{O}_X((d + m)C_0 + (r + n)f_0) \otimes \pi^*(L_2 \otimes N_0) \to E \to \quad (5.2)$$

$$\to J_Y \otimes \mathcal{O}_X((d' - m)C_0 + (\beta - r - n)f_0) \otimes \pi^*(L_2^{-1} \otimes N_0^{-1} \otimes L_0) \to 0 ,$$

where $L_0 \in Pic_0(C)$ is given by

$$\bigwedge^2 E = \mathcal{O}_X(\alpha C_0 + \beta f_0) \otimes \pi^*(L_0) .$$

For a fibre $f$ of $\pi$ such that $f \cap Y = \emptyset$ we get from (5.2) the exact sequence

$$0 \to \mathcal{O}_f(d + m) \to E|_f \to \mathcal{O}_f(d' - m) \to 0 .$$

But $H^1(f, \mathcal{O}_f(d - d' + 2m)) = 0$ since $d - d' + 2m \geq 0$, hence

$$E|_f \cong \mathcal{O}_f(d + m) \oplus \mathcal{O}_f(d' - m).$$

It follows $m = 0$, hence $n \geq 0$ and $A$ is effective. Again, from (5.2) we get the injective map

$$0 \to \mathcal{O}_X \to E(-dC_0) \otimes \mathcal{O}_X((-r-n)f_0) \otimes \pi^*(L_2^{-1} \otimes N_0^{-1}),$$

hence

$$H^0(X, E(-dC_0) \otimes \mathcal{O}_X((-r-n)f_0) \otimes \pi^*(L_2^{-1} \otimes N_0^{-1})) \neq 0.$$

It follows, by the definition of $r$, that $-r - n \geq -r$, hence $n = 0$. From $m = 0$, $n = 0$ and $A$ effective we get $Z = 0$ and $N_0 \cong \mathcal{O}_C$.

Suppose that $\alpha, \beta, \gamma$ are fixed; one may ask for which integers $d$ and $r$ is $M(\alpha, \beta, \gamma, d, r)$ nonempty ? Of course, we have the necessary conditions

$$2d \geq \alpha, \ \gamma + \alpha(de - r) - \beta d + 2dr - d^2 e \geq 0.$$

We shall give the following sufficient conditions:

**Proposition 5.49** *If $2d > \alpha$ and $\gamma + \alpha(de - r) - \beta d + 2dr - d^2 e \geq 0$, then the set $M(\alpha, \beta, \gamma, d, r)$ is nonempty.*

*Proof. Step 1.* We shall prove firstly that every algebraic 2-vector bundle given by an extension (5.1) belongs to $M(\alpha, \beta, \gamma, d, r)$.

Obviously, an algebraic 2-vector bundle given by such an extension has numerical Chern classes $c_1 = (\alpha, \beta)$ and $c_2 = \gamma$. By restricting the exact sequence above to a fibre $f$, provided that $f \cap Y = \emptyset$, we get that the splitting type of $E$ is $(d, d')$ (one has the exact sequence

$$0 \to \mathcal{O}_f(d) \to E|_f \to \mathcal{O}_f(d') \to 0$$

and $H^1(f, \mathcal{O}_F(d - d')) = 0$ since $d \geq d'$. Put the exact sequence (5.1) under the equivalent form

$$0 \to \mathcal{O}_X \to E(-dC_0) \otimes \pi^*(\check{L}) \to \qquad\qquad (5.3)$$

$$\to J_Y \otimes \mathcal{O}_X((d' - d)C_0 + (s - r)f_0) \otimes \pi^*(L_1 \otimes L_2^{-1}) \to 0,$$

where $\check{L} = \mathcal{O}_C(-r) \otimes L_2^{-1} \in \mathrm{Pic}(C)$ and $\deg(\check{L}) = -r$. From the long exact cohomology sequence associated to (5.3) we get the injective map

$$0 \to H^0(X, \mathcal{O}_X) \to H^0(X, E(-dC_0) \otimes \pi^*(\check{L})),$$

hence $H^0(X, E(-dC_0) \otimes \pi^*(\check{L})) \neq 0$. In order to prove that $r$ is the second numerical invariant of $E$ we must show that

$$H^0(X, E(-dC_0) \otimes \pi^*(\check{L} \otimes L')) = 0$$

for any $L' \in \mathrm{Pic}(C)$ with $\deg(L') < 0$. Tensoring (5.3) by $\pi^*(L')$ one gets

$$0 \to H^0(X, \pi^*(L')) \to H^0(X, E(-dC_0) \otimes \pi^*(\check{L} \otimes L')) \to$$

$$\to H^0(X, J_Y \otimes \mathcal{O}_X((d' - d)C_0 + (s - r)f_0) \otimes \pi^*(L_1 \otimes L_2^{-1} \otimes L')) \to \cdots.$$

By the projection formula, we have

$$H^0(X, \mathcal{O}_X((d' - d)C_0 + (s - r)f_0) \otimes \pi^*(L_1 \otimes L_2^{-1} \otimes L')) \cong$$

$$\cong H^0(C, \pi_*(\mathcal{O}_X((d' - d)C_0)) \otimes \mathcal{O}_C(s - r) \otimes L_1 \otimes L_2^{-1} \otimes L') = 0$$

$(\pi_*(\mathcal{O}_X((d' - d)C_0)) = 0$ as $d' - d < 0$). Therefore, the third term in the last exact sequence is zero too, hence

$$H^0(X, E(-dC_0) \otimes \pi^*(\tilde{L} \otimes L')) \cong H^0(X, \pi^*(L')) \cong H^0(C, L') = 0$$

$(\deg(L') < 0)$ and we conclude.

*Step 2.* For every fixed data $Y$, $L_1$ $L_2$, where $Y \subset X$ is a locally complete intersection of codimension 2 with

$$\deg(Y) = \gamma + \alpha(de - r) - \beta d + 2dr - d^2 e$$

(or the empty set) and $L_1$, $L_2 \in \mathrm{Pic}_0(C)$, there exist bundles appearing as extensions (5.1).

Let us denote

$$N_1 = \mathcal{O}_X(d'C_0 + sf_0) \otimes \pi^*(L_1)$$

and

$$N_2 = \mathcal{O}_X(dC_0 + rf_0) \otimes \pi^*(L_2) .$$

Consider the spectral sequence of term

$$E_2^{p,q} = H^p(X, \mathcal{E}xt_X^q(J_Y \otimes N_1, N_2)) ,$$

which converges to $E^{p+q} = \mathrm{Ext}^{p+q}(J_Y \otimes N_1, N_2)$. We have

$$\mathcal{E}xt^0(J_Y \otimes N_1, N_2) \cong N_2 \otimes N_1^{-1}$$

and

$$\mathcal{E}xt^1(J_Y \otimes N_1, N_2) \cong \mathcal{E}xt^2(\mathcal{O}_Y \otimes N_1, N_2) \cong \mathcal{O}_Y$$

(for the last isomorphism we use the connection between the dualising sheaves $\omega_Y$ and $\omega_X$ and the fact that $\dim(Y) = 0$). But

$$H^2(X, N_2 \otimes N_1^{-1}) \cong H^2(C, \pi_*(\mathcal{O}_X((d - d')C_0)) \otimes \mathcal{O}_C(r - s) \otimes L_2 \otimes L_1^{-1}) = 0$$

(see [Hh1], p. 371). It follows that the exact sequence of lower degree terms becomes:

$$0 \to H^1(X, N_2 \otimes N_1^{-1}) \to \mathrm{Ext}^1(J_Y \otimes N_1, N_2) \to H^0(Y, \mathcal{O}_Y) \to 0 .$$

Now, by the Serre method (Theorem 4.3), any element belonging to

$$\mathrm{Ext}^1(J_Y \otimes N_1, N_2)$$

which has an invertible image in $H^0(Y, \mathcal{O}_Y)$, defines an extension of the desired form with $E$ a 2-vector bundle (if $\deg(Y) > 0$). If $\deg(Y) = 0$, then

$$\mathrm{Ext}^1(N_1, N_2) \cong H^1(X, N_2 \otimes N_1^{-1})$$

and there is at least one such vector bundle (the direct sum $N_1 \oplus N_2$). It follows that the set $M(\alpha, \beta, \gamma, d, r)$ is nonempty.

*Remark.* Observe that the inequalities from the proposition are satisfied for an infinite numbers of pairs $(d, r)$.

Let $H$ be an ample divisor on $X$ and let $E$ be an algebraic 2-vector bundle over $X$ with fixed numerical Chern classes $c_1 = (\alpha, \beta)$, $c_2 = \gamma$. Suppose that $E$ is $H$-stable; one may ask if the numerical invariants $d$ and $r$ are determined by $\alpha, \beta, \gamma$ ? Generally, the invariants $d$ and $r$ are not determined by $\alpha, \beta, \gamma$ if $E$ is $H$-stable. In [H-Sp] Hoppe and Spindler studied $H$-stable 2-vector bundles over ruled surfaces by introducing another two numerical invariants (which can be related with $d$ and $r$) and proved that a finite set of pairs of these invariants appears, each of them corresponding to a stratum in the moduli space. In fact, these invariants ($d$ and $r$) are not constant in a flat family, while the numerical (topological) Chern classes $c_1$ and $c_2$ are constant. Note also, the following result of Takemoto ( [Tk]): For some 2-vector bundle $E$ over a ruled surface $X$, which is $H$-stable, there exists an ample divisor $H'$ on $X$ such that $E$ is not $H'$-stable.

*Remark.*   Some of the sets $M(\alpha, \beta, \gamma, d, r)$ appear as the sets of vector bundles $E_\zeta(c_1, c_2)$ defined by Qin in [Q3].

We shall prove that the set $M(\alpha, \beta, \gamma, d, r)$ is bounded. Then this set has a quotient structure of algebraic variety (cf. [Kl]).

A set $M$ of vector bundles over a $\mathbb{C}$-scheme $X$ is called *bounded* if there exist an algebraic $\mathbb{C}$-scheme $T$ and a vector bundle $V$ over $T \times X$ such that every $E \in M$ is isomorphic with $V_t = V|_{t \times X}$, for some closed point $t \in T$.

We need some preliminary results.

**Lemma 5.50** *With the above notations, one has:*

$$\dim \ H^0(X, \mathcal{O}_X(dC_0 + rf_0) \otimes \pi^*(L_2)) \leq m_0' \ ,$$

*where $m_0'$ does not depend on $L_2 \in Pic_0(C)$, but depends on $d$, $r$ and $X$ (hence it is fixed).*

*Proof.*  Let us denote $L = \mathcal{O}_C(rp_0) \otimes L_2$; then $\deg(L) = r$. For $d < 0$ we have

$$H^0(X, \mathcal{O}_X(dC_0) \otimes \pi^*(L)) \cong H^0(C, \pi_*(\mathcal{O}_X(dC_0)) \otimes L) = 0 \ ,$$

since $\pi_*(\mathcal{O}_X(dC_0)) = 0$ (see [Hh1], p. 253); hence we take $m_0' = 0$.

If $d \geq 0$ we get (see [Hh1], p. 253):

$$H^0(X, \mathcal{O}_X(dC_0) \otimes \pi^*(L)) \cong H^0(C, \pi_*(\mathcal{O}_X(dC_0)) \otimes L) \cong H^0(C, S^d \mathcal{E} \otimes L) \ ,$$

where $\mathcal{E} \cong \pi_*(\mathcal{O}_X(C_0))$ is the normalized 2-vector bundle over $C$, which defines the ruled surface $X$.

We have the following facts:

If $C$ is an algebraic curve and $V$ is a fixed vector bundle over $C$, then:

(1) There exists an integer $q_0$ such that $\dim H^0(C, V \otimes L) \leq q_0$, $\forall L \in Pic_r(C)$.

(2) There exists an integer $q_1$ such that $\dim H^1(C, V \otimes L) \leq q_1$, $\forall\, L \in \mathrm{Pic}_r(C)$, where $\mathrm{Pic}_r(C)$ is the scheme of line bundles on $C$ of degree $r$.

Indeed, consider on $\mathrm{Pic}_r(C) \times C$ the universal Poincaré bundle $\mathcal{P}$ and take the sheaves $F^i = R^i p_{1*}(p_2^* V \otimes \mathcal{P})$, where $i = 0, 1$ and $p_1$, $p_2$ are the canonical projections. The sheaves $F^i$ are coherent and, for any closed point $L \in \mathrm{Pic}_r(C)$, we have

$$F^i_L \otimes k(L) \cong H^i(C, V \otimes L) \,.$$

Since the functions

$$L \mapsto \dim_{k(L)} F^i_L \otimes k(L)$$

are upper-semicontinuous (see [B-S], p. 134), the assertions (1) and (2) follow by a standard argument.

Applying the fact (1) for $V = S^d \mathcal{E}$, the proof of the lemma is over.

Let $H_1 = C_0 + bf_0$ be an ample divisor on $X$ ($b > \max\{e, 0\}$; see [Hh1], p. 382). Then, there exists an integer $n_0 > 0$ such that the sheaf $\mathcal{O}_X(n_0 H_1)$ is generated by global sections. Let us denote $H = n_0 H_1$.

**Lemma 5.51** *With the notations above one has;*

$$\dim H^1(X, \mathcal{O}_X((d - n_0)C_0 + (r - n_0 b)f_0) \otimes \pi^*(L_2)) \leq m_1' \,,$$

*where $m_1'$ does not depend on $L_2 \in \mathrm{Pic}_0(C)$, but depends on $d - n_0$, $r - n_0 b$ and $X$ (hence it is fixed).*

*Proof.* Let us denote $L = \mathcal{O}_C((r - n_0 b)p_0) \otimes L_2$; then $\deg(L) = r - n_0 b$. If $d - n_0 \geq 0$, by [Hh1], p. 253, p. 371 and by projection formula, one gets

$$H^1(X, \mathcal{O}_X((d - n_0)C_0) \otimes \pi^*(L)) \cong H^1(C, (S^{d-n_0}\mathcal{E}) \otimes L) \,.$$

By the fact (2) above (for $V = S^{d-n_0}\mathcal{E}$) we get the boundness with an integer depending only on $d - n_0$, $r - n_0 b$ and $X$.

If $d - n_0 = -1$, again by [Hh1], p. 253 and the projection formula one has

$$R^1 \pi_*(\mathcal{O}_X(-C_0) \otimes \pi^*(L)) = R^1 \pi_*(\mathcal{O}_X(-C_0)) \otimes L = 0 \,.$$

But

$$R^i \pi_*(\mathcal{O}_X(-C_0) \otimes \pi^*(L)) = 0$$

for $i \geq 2$ (cf. [Hh1], p. 253) hence, it follows

$$H^1(X, \mathcal{O}_X(-C_0) \otimes \pi^*(L)) \cong H^1(C, \pi_*(\mathcal{O}_X(-C_0)) \otimes L) = 0 \,.$$

If $d - n_0 \leq -2$, we apply Serre's duality on $X$:

$$H^1(X, \mathcal{O}_X((d - n_0)C_0) \otimes \pi^*(L)) \cong H^1(X, \mathcal{O}_X((n_0 - d - 2)C_0) \otimes \pi^*(L')) \,,$$

where $L' \in \mathrm{Pic}(C)$ and $\deg(L') = n_0 b - r + 2g - 2 - e$. But $n_0 - d - 2 \geq 0$ and the boundness follows as in the first case of the proof of this lemma.

**Theorem 5.52** *The set $M(\alpha, \beta, \gamma, d, r)$ is bounded.*

*Proof.* By the Theorem 1.13 in [Kl], it suffices to verify the following conditions:

(a) The set of Hilbert polynomials $\chi(E(nH))$, when $E \in M$, is finite.

(b) There exist integers $m_0$ and $m_1$ such that for every $E$ in $M$ one has

$$\dim H^0(X, E) \leq m_0 \quad \text{and} \quad \dim H^0(C', E|_{C'}) \leq m_1 \,,$$

for $C' \in |H|$ a generic member of the linear system $|H|$.

In fact, we should verify also the condition

$$\dim H^0(C' \cap C'', E|_{C' \cap C''}) \leq m_2 \,,$$

for $C'$, $C'' \in |H|$ generic and transversal. But $C' \cap C''$ is a finite set of points of cardinality $(C')^2$ (which is fixed) and we can take $m_2 = 2(C')^2$.

By Riemann-Roch Theorem for the vector bundle $E$ over $X$ it follows that only one Hilbert polynomial appears, hence (a) is fulfilled.

We denote, as usual,

$$h^0(X, F) = \dim H^0(X, F)$$

for any coherent sheaf on $X$. From the long exact sequence of cohomology associated to (5.1) we get the inequality

$$h^0(X, E) \leq h^0(X, \mathcal{O}_X(dC_0 + rf_0) \otimes \pi^*(L_2)) + \tag{5.4}$$

$$+ h^0(X, \mathcal{O}_X(d'C_0 + sf_0) \otimes \pi^*(L_1)) \,,$$

because $h^0(X, J_Y \otimes F) \leq h^0(X, F)$ for a locally free sheaf $F$. By Lemma 5.50 and the inequality (5.4) one gets $h^0(X, E) \leq m_0$ for every $E \in M$, where the integer $m_0$ depends only on $d, r$ and on $X$ (so, it is fixed).

By Bertini Theorem one has that the generic $C' \in |H|$ is a nonsingular curve. From the exact sequence

$$0 \to E(-H) \to E \to E|_{C'} \to 0 \,, \tag{5.5}$$

we obtain the long exact sequence of cohomology

$$0 \to H^0(X, E(-H)) \to H^0(X, E) \to H^0(C', E|_{C'}) \to H^1(X, E(-H)) \to \cdots .$$

It follows the inequality

$$h^0(C', E|_{C'}) \leq h^0(X, E) + h^1(X, E(-H)) \,, \tag{5.6}$$

and thus it suffices to show that $h^1(X, E(-H))$ is bounded. By tensoring the exact sequence (5.1) with $\mathcal{O}_X(-H)$ and by taking the cohomology we get, as above, the inequality

$$h^1(X, E(-H)) \leq h^1(X, \mathcal{O}_X((d - n_0)C_0 + (r - n_0 b)f_0) \otimes \pi^*(L_2)) + \tag{5.7}$$

$$+ h^1(X, J_Y \otimes \mathcal{O}_X((d' - n_0)C_0 + (s - n_0 b)f_0) \otimes \pi^*(L_1)) \,.$$

Also, we have

$$h^1(X, J_Y \otimes \mathcal{O}_X((d' - n_0)C_0 + (s - n_0 b)f_0) \otimes \pi^*(L_1)) \leq h^0(Y, \mathcal{O}_Y) + \tag{5.8}$$

$$+h^1(X, \mathcal{O}_X((d' - n_0)C_0 + (s - n_0 b)f_0) \otimes \pi^*(L_1)) \ .$$

By Lemma 5.51, the inequalities (5.7), (5.8) and by the fact that $h^0(Y, \mathcal{O}_Y) = \deg(Y)$ is fixed, it follows that the condition (b) is fulfilled and the proof is over.

For the following result see also [B-S1, B-S2]:

**Corollary 5.53** *The set $M(d, r)$ has a structure of algebraic variety.*

By using the variation of global Ext in a deformation (see [B-P-S]) one obtains the following results (for simplicity, we restrict to the case $c_1 = 0$, $c_2 = 0$); see [B-S1, B-S2]:

**Theorem 5.54** *Suppose $d > 0$. Then:*

*(1) The set $M(d, r)$ carries a natural structure of algebraic variety.*

*(2) There exists, locally relative to $M(d, r)$, a tautological bundle (i.e. for every affine open subset $V$ of $M(d, r)$ there is a bundle $F$ over $V \times X$ such that for each $t \in V$ there is an isomorphism $F/\boldsymbol{m}_t F \cong E_t$, where $E_t$ denotes a 2-vector bundle corresponding to the class $t$).*

In the case of rational ruled surfaces we get more precise results:

**Theorem 5.55** *Assume $C = \mathbb{P}^1$ (i.e. $X \cong \Sigma_e$, $e \geq 0$ is a rational ruled surface). Then the set $M(d, r)$ is a nonsingular, connected, quasi-projective, rational variety of dimension indicated below:*

*(a) when $de = 2r$ and $e \geq 1$ (uniform bundles)*
$$\dim M(d, r) = d(de + e - 2)/2 \ ;$$

*(b) when $e = 0$ ($r = 0$) (uniform bundles)*
$$\dim M(d, 0) = -1 \quad (M = \emptyset) \ ;$$

*(c) when $2r - de > 0$ and $e \geq 1$ (non-uniform bundles), there are two posibilities:*
$$r > de, \quad \dim M(d, r) = 3d(2r - de) - 1 \ ,$$

*or*
$$r \leq de, \quad \dim M(d, r) = (2d - s + 1)(es + 2de - 4r - 2)/2 + 3d(2r - de) - 1 \ ,$$
*where $s = [(2r + 1)/e] + 1$;*

*(d) when $2r - de > 0$ but $e = 0$ (non-uniform bundles)*
$$\dim M(d, r) = 6dr - 1 \ .$$

**Theorem 5.56** *Let $X$ be the rational ruled surface $\Sigma_1$ ($e = 1$). There is not globally a tautological bundle relative to $M(1, 1)$.*

# References

[A-K]   Altman, A.B., Kleiman,S.L.: Compactifying the Picard scheme, Adv. Math. **35**, (1980) 50–112

[An]    Anghel, C.: Une construction de fibrés vectoriels stables de rang deux avec $c_2$ suffisamment grand sur une surface algébrique, C.R. Acad. Sci. Paris, t. **318**, Serie I (1994) 541–542

[At1]   Atiyah, M.F.: Complex analytic connections in fibre bundles, Trans. Amer. Math. Soc. **85**, (1957) 181–207

[At2]   Atiyah, M.F.: Vector bundles over an elliptic curve, Proc. London Math. Soc. **7**, (1957) 414–452

[A-R]   Atiyah, M.F., Rees, E.: Vector bundles on projective 3-space, Invent. Math. **35**, (1976) 131–153

[Bd]    Bădescu, L.: Suprafete algebrice, Edit. Acad. Romania, Bucharest 1981

[Bl1]   Ballico, E.: A problem list on vector bundles, In: Complex Analysis and Geometry, Eds: V. Ancona, A. Silva, Plenum Press, New-York 1993, 387–402

[Bl2]   Ballico, E.: On the components of moduli of simple sheaves on a surface, Forum Math. **6**, (1994) 35–41

[Bn]    Bănică, C.: Topologisch triviale holomorphe Vektorbündel auf $\mathbb{P}^n$, J. reine angew. Math. **344**, (1983) 102–119

[B-C]   Bănică, C., Coandă, J.: Existence of rank 3 vector bundles with given Chern classes on homogeneous rational 3-folds, Manuscripta Math. **51**, (1985) 121–143

[B-L]   Bănică, C., Le Potier, J.: Sur l'existence des fibrés vectoriels holomorphes sur les surfaces non-algébriques, J. reine angew. Math. **378**, (1987) 1–31

[B-P]   Bănică, C., Putinar, M.: On complex vector bundles on projective threefolds, Invent. Math. **88**, (1987) 427–438

[B-P-S] Bănică, C., Putinar, M., Schumacher, G.: Variation der globalen Ext in Deformationen kompakter komplexer Räume, Math. Ann. **250**, (1980) 135–155

[B-S]   Bănică, C., Stănăsilă, O.: Algebraic Methods in the global theory of complex spaces, John Wiley & Sons, New York 1976

[B-St]  Bănică, C., Stoia, M.: Holomorphic vector bundles, In: Analiza Complexa, Ed. St. Encicl., Bucharest 1988, 223–262

[Ba]    Baran, A.: Duality for the hyperext on complex spaces, Preprint Increst **15**, Bucharest (1985)

[Bt]    Barth, W.: Moduli of vector bundles on projective plane, Invent. Math. **42**, (1977) 63–91

[B-P-V] Barth, W., Peters, C., Van de Ven, A.: Compact Complex Surfaces, Ergebnisse der Mathematik und ihrer Grenzgebite, 3. Folge, Band 4, Springer, Berlin Heidelberg New York Tokyo 1984

[Bu1]   Bauer, S.: Some nonreduced moduli of bundles and Donaldson invariants for Dolgachev surfaces, J. reine angew. Math. **424**, (1992) 149–180

[Bu2]    Bauer, S.: Diffeomorphism classification of elliptic surfaces with $p_g = 1$, J. reine angew. Math. **451**, (1994) 89–148

[Be]    Beauville, A.: Surfaces algébriques complexes, Astérisque **54**, Soc. Math. France, Paris 1978

[Bo1]    Bogomolov, F.A.: Classification of surfaces of class $VII_0$ with $b_2 = 0$, Math. USSR-Izv. vol. **10**(2), (1976) 255–269

[Bo2]    Bogomolov, F.A.: Holomorphic tensors and vector bundles on projective varieties, Izv. Akad. Nauk SSSR, **42**, (1978) 499–555

[B-H]    Bombieri, E., Husemoller, D.: Classification and embeddings of surfaces, In: Algebraic Geometry-Arcata 1974, Amer. Math. Soc. Proc. Symp. Pure Math. **29**, (1975) 329–420

[Br-Sr]    Borel, A., Serre, J.P.: Le théorème de Riemann-Roch (d'après Grothendieck), Bull. Soc. Math. France **86**, (1958) 97–136

[Bm-Hu]Braam, P., Hurtubise, J.: Instantons on Hopf surfaces and monopoles on solid tori, J. reine angew. Math. **400**, (1989) 146–172

[B-T-T1]O'Brian, N.R., Toledo, D., Tong, Y.L.L.: Hirzebruch-Riemann-Roch for coherent sheaves, Amer. J. Math. **103**, (1981) 253–271

[B-T-T2]O'Brian, N.R., Toledo, D., Tong, Y.L.L.: Grothendieck-Riemann-Roch for complex manifolds, Bull. Amer. Math. Soc. **5**, (1981) 182–184

[Br1]    Brînzănescu, V.: Algebraic 2-vector bundles on ruled surfaces, Ann. Univ. Ferrara-Sez VII-Sc. Mat. Vol. XXXVII, (1991) 55–64

[Br2]    Brînzănescu, V.: Neron-Severi group for nonalgebraic elliptic surfaces I: elliptic bundle case, Manuscripta Math. **79**, (1993) 187–195

[Br3]    Brînzănescu, V.: The Picard group of a primary Kodaira surface, Math. Ann. **296**, (1993) 725–738

[Br4]    Brînzănescu, V.: A simple proof of a Bogomolov type inequality in the case of nonalgebraic surfaces, Rev. Roumaine Math. Pures Appl. **38**(7-8), (1993) 631–633

[Br5]    Brînzănescu, V.: Neron-Severi group for nonalgebraic elliptic surfaces II: non-kählerian case, Manuscripta Math. **84**, (1994) 415–420

[B-F1]    Brînzănescu, V., Flondor, P.: Holomorphic 2-vector bundles on nonalgebraic 2-tori, J. reine angew. Math. **363**, (1985) 47–58

[B-F2]    Brînzănescu, V., Flondor, P.: Simple filtrable 2-vector bundles on complex surfaces without divisors, Rev. Roumaine Math. Pures Appl. **31**, (1986) 507–512

[B-F3]    Brînzănescu, V., Flondor, P.: Quadratic intersection form and 2-vector bundles on nonalgebraic surfaces, In: Proc. Inter. Conf. Alg. Geom., Berlin 1985, Teubner, Leipzig 1986

[B-F4]    Brînzănescu, V., Flondor, P.: Vector bundles on nonalgebraic surfaces, In: Analiza Complexa, Ed. St. Encicl., Bucharest 1988, 263–294

[B-S1]    Brînzănescu, V., Stoia, M.: Topologically trivial algebraic 2-vector bundles on ruled surfaces I, Rev. Roumaine Math. Pures Appl. **29**, (1984) 661–673

[B-S2]    Brînzănescu, V., Stoia, M.: Topologically trivial algebraic 2-vector bundles on ruled surfaces II, In: Lect. Notes Math. **1056**, Springer, Berlin Heidelberg New York 1984

[B-U]    Brînzănescu, V., Ueno, K.: Neron-Severi group for torus quasi-bundles over curves, Preprint MPI **60** (1994)

[Bs]    Brosius, J.E.: Rank-2 vector bundles on a ruled surface I, Math. Ann. **265**, (1983) 155–168, II, Math. Ann. **266**, (1983) 199–214

[Bh]    Buchdahl, N.: Hermitian-Einstein connections and stable vector bundles over compact complex surfaces, Math. Ann. **280**, (1988) 625–648

[Bc]    Bucur, I.: Sur une extension d'un théorème de Wu, Rev. Roumaine Math. Pures Appl. **3**(3), (1958) 427–428

[Cr]  Cartan, H.: Quotient d'un espace analytique par un groupe d'automorphismes, In: Algebraic Geometry and Topology, Princeton Univ. Press, Princeton 1957, 90–102

[Ct]  Catanese, F.: On the moduli spaces of surfaces of general type, J. Differ. Geom. **19**, (1984) 483–515

[C-K]  Chow, W.L., Kodaira, K.: On analytic surfaces with two independent meromorphic functions, Proc. Nat. Acad. Sci. U.S.A. **38**, (1952) 319–325

[Cd]  Coandă, J.: The Chern classes of the stable rank 3 vector bundles on $\mathbb{P}^3$, Math. Ann. **273**, (1985) 65–79

[Do1]  Donaldson, S.K.: A new proof of a theorem of Narasimhan and Seshadri, J. Differ. Geom. **18**, (1983) 269–277

[Do2]  Donaldson, S.K.: Anti-self-dual Yang-Mills connections over complex algebraic surfaces and stable vector bundles, Proc. London Math. Soc. **50**, (1985) 1–26

[Do3]  Donaldson, S.K.: Polynomial invariants for smooth 4-manifolds, Topology **29**, (1990) 257–315

[D-K]  Donaldson, S.K., Kronheimer, P.B.: The Geometry of Four-Manifolds, Clarendon, Oxford 1990

[Du]  Douady, A.: Le probleme des modules pour les sous-espaces analytiques compacts d'un espace analytique donné, Ann. Inst. Fourier **16**(1), (1966) 1–95

[Dz1]  Drezet, J.: Fibrés exceptionnels et variétés de modules de faisceaux semi-stables sur $\mathbb{P}_2$, J. reine angew. Math. **380**, (1987) 14–58

[Dz2]  Drezet, J.: Cohomologie des variétés de modules de hauteur nulle, Math. Ann. **281**, (1988) 43–85

[Dz3]  Drezet, J.: Groupe de Picard des variétés de modules de faisceaux semi-stables sur $\mathbb{P}_2(\mathbb{C})$, Ann. Inst. Fourier **38**(3), (1988) 105–168

[D-L]  Drezet, J., Le Potier, J.: Fibrés stables et fibrés exceptionnels sur $\mathbb{P}_2$, Ann. Sc. Ec. Norm. Sup. **18**, (1985) 193–244

[E-F]  Elencwajg, G., Forster, O.: Vector bundles on manifolds without divisors and a theorem of deformation, Ann. Inst. Fourier **32**(4), (1982) 25–51

[Ek]  Enoki, I.: Surfaces of class $VII_0$ with curves, Tôhoku Math. J. **33**, (1981) 453–492

[En]  Enriques, F.: Le superficie algebriche, Zanichelli, Bologna 1949

[Fh]  Fischer, G.: Complex analytic geometry, Lect. Notes Math. **538**, Springer, Berlin Heidelberg 1976

[F-F]  Fischer, G., Forster, O.: Ein Endlichkeitssatz für Hyperflächen auf kompakter komplexen Räumen, J. reine angew. Math. **306**, (1969) 88–93

[Fo]  Forster, O.: Riemann surfaces, Springer, Berlin Heidelberg New York 1982

[F-K]  Forster, O., Knorr, K.: Über die Deformationen von Vektorraumbündeln auf kompakten komplexen Räumen, Math. Ann. **209**, (1974) 291–346

[Fr]  Friedman, R.: Rank two vector bundles over regular elliptic surfaces, Invent. Math. **96**, (1989) 283–332

[F-M1]  Friedman, R., Morgan, J.W.: On the diffeomorphism type of certain algebraic surfaces I, J. Differ. Geom. **27**, (1988) 297–369, II, J. Differ. Geom. **27**, (1988) 371–398

[F-M2]  Friedman, R., Morgan, J.W.: Algebraic surfaces and 4-manifolds: some conjectures and speculations, Bull. Amer. Math. Soc. **18**, (1988) 1–19

[F-M3]  Friedman, R., Morgan, J.W.: Smooth Four-Manifolds and Complex Surfaces, Ergebnisse der Mathematik und ihrer Grenzgebite, 3. Folge, Band 27, Springer, Berlin Heidelberg New York London Paris Tokyo Hong Kong Barcelona Budapest 1994

[F-S]  Fujiki, A., Schumacher, G.: The moduli space of Hermite-Einstein bundles on a compact Kähler manifold, Proc. Japan Acad. **63**, (1987) 69–72

[Ga]  Gauduchon, P.: Sur la 1-forme de torsion d'une variété hermitienne compacte, Math. Ann. **267**, (1984) 495–518

[Gi1]   Gieseker, D.: On moduli of vector bundles on an algebraic surface, Ann. Math. **106**, (1977) 45–60

[Gi2]   Gieseker, D.: Global moduli for surfaces of general type, Invent. Math. **43**, (1977) 233–282

[Gi3]   Gieseker, D.: A construction of stable bundles on an algebraic surface, J. Differ. Geom. **27**, (1988) 137–154

[G-L]   Gieseker, D., Li, J.: Irreducibility of moduli of rank-2 vector bundles on algebraic surfaces, J. Differ. Geom. **40**, (1994) 23–104

[Go]    Godement, R.: Topologie Algébrique et Théorie des Faisceaux, Act. Sci. Ind. 1252, Hermann, Paris 1958

[Gt]    Göttsche, L.: Change of polarization and Hodge numbers of moduli spaces of torsion free sheaves on surfaces, Preprint MPI **9** (1994)

[Gt-H]  Göttsche, L., Huybrechts, D.: Hodge numbers of moduli spaces of stable bundles on K3 surfaces, Preprint MPI **80** (1994)

[Gd1]   O'Grady, K.: Donaldson's polynomials for K3 surfaces, J. Differ. Geom. **35**, (1992) 415–427

[Gd2]   O'Grady, K.: Moduli of vector bundles on projective surfaces: some basic results, Preprint 1994

[Gr1]   Grauert, H.: Analytische Faserungen über holomorph-vollständigen Räumen, Math. Ann. **133**, (1958) 263–273

[Gr2]   Grauert, H.: Über Modifikationen und exzeptionelle analytische Mengen, Math. Ann. **146**, (1962) 331–368

[G-R1]  Grauert, H., Remmert, R.: Analytische Stellenalgebren, Springer 1971

[G-R2]  Grauert, H., Remmert, R.: Theorie der Steinschen Räume, Springer 1977

[G-R3]  Grauert, H., Remmert, R.: Coherent analytic sheaves, Springer, Berlin Heidelberg New York 1984

[Gf]    Griffiths, Ph.: Topics in algebraic and analytic geometry, Princeton 1974

[G-H]   Griffiths, Ph., Harris, J.: Principles of Algebraic Geometry, John Wiley & Sons 1978

[Gk1]   Grothendieck, A.: Sur la classification des fibrés holomorphes sur la sphère de Riemann, Amer. J. Math. **79**, (1957) 121–138

[Gk2]   Grothendieck, A.: Sur quelques points d'algèbre homologique, Tôhoku Math. J. **9**, (1957) 119–221

[Gk3]   Grothendieck, A.: La théorie des classes de Chern, Bull. Soc. Math. France **86**, (1958) 137–154

[Gk4]   Grothendieck, A.: Techniques de construction en géométrie analytique, Sém. H. Cartan 1960/1961,Exposés 13, 16

[Gk5]   Grothendieck, A.: Techniques de construction et théorèmes d'existence en géometrie algébrique, IV: Les schéma de Hilbert, Sém. Bourbaki, no.221, 1961

[H-N]   Harder, G., Narasimhan, M.S.: On the cohomology groups of moduli spaces of vector bundles over curves, Math. Ann. **212**, (1975) 215–248

[Hh1]   Hartshorne, R.: Algebraic Geometry, Graduate Texts in Math. 49, Springer, Berlin Heidelberg 1977

[Hh2]   Hartshorne, R.: Stable reflexive sheaves, Math. Ann. **254**, (1980) 121–176

[H-L]   Hirschowitz, A., Laszlo, Y.: A propos de l'existence de fibrés stable sur les surfaces, J. reine angew. Math. **460**, (1995) 55–68

[Hr]    Hirzebruch, F.: Topological methods in algebraic geometry, Grundl. Math. Wiss. **131**, Springer, Berlin Heidelberg 1966

[H-V]   Hirzebruch, F., Van de Ven, A.: Hilbert modular surfaces and the classification of algebraic surfaces, Invent. Math. **23**, (1974) 1–29

[H-S]    Hochschild, G., Serre, J.P.: Cohomology of group extensions, Trans. Am. Math. Soc. 74, (1953) 110–134

[Hf]     Höfer, T.: Remarks on torus principal bundles, J. Math. Kyoto Univ. (JMKYAZ) 33(1), (1993) 227–259

[H-Sp]   Hoppe, H.J., Spindler, H.: Modulräume stabile 2-Bündel auf Regelflächen, Math. Ann. 249, (1980) 127–140

[Ho]     Horrocks, G.: A construction for locally free sheaves, Topology 7, (1968) 117–120

[Hu1]    Hulek, K.: Stable rank-2 vector bundles on $\mathbb{P}_2$ with $c_1$ odd, Math. Ann. 242, (1979) 241–266

[Hu2]    Hulek, K.: On the classification of stable rank $r$ vector bundles over the projective plane, Progress in Math. 7, Vector Bundles and Differential Equations, Birkhäuser 1980, 113–114

[Hs]     Husemoller, D.: Fibre Bundles, McGraw-Hill 1966

[Hy]     Huybrechts, D.: Complete curves in moduli spaces of stable bundles on surfaces Math. Ann. 298, (1994) 67–78

[In]     Inoue, M.: On surfaces of class $VII_0$, Invent. Math. 24, (1974) 269–310

[Ka]     Kato, M.: Compact complex manifolds containing "global" spherical shells I, In: Int. Symp. Alg. Geom., Kyoto 1977, Kinokuniya, Tokyo 1978, 45–84

[Ke]     Kempf, G.: Complex abelian varieties and theta functions, Springer, Berlin Heidelberg New York 1991

[Ki]     Kim, H.J.: Curvature and holomorphic vector bundles, Ph.D. thesis, Berkeley 1985

[Kl]     Kleiman, S.: Les théorèmes de finitude pour les Foncteurs de Picard, In: Théories des intersections et théorème de Riemann-Roch, SGA VI, Exp.XIII, Lect. Notes Math. 225, Springer 1971, 616–666

[Ko1]    Kobayashi, S.: First Chern class and holomorphic tensor fields, Nagoya Math. J. 77, (1980) 5–11

[Ko2]    Kobayashi, S.: Curvature and stability of vector bundles, Proc. Japan Acad. 58, (1982) 158–162

[Ko3]    Kobayashi, S.: Differential geometry of complex vector bundles, Iwanami Shoten and Princeton Univ. Press 1987

[Kd1]    Kodaira, K.: On compact complex analytic surfaces I, Ann. Math. 71, (1960) 111–152, II, Ann. Math. 77, (1963) 563–626, III, Ann. Math. 78, (1963) 1–40

[Kd2]    Kodaira, K.: On the structure of compact complex analytic surfaces I, Amer. J. Math. 86, (1964) 751–798, II, Amer. J. Math. 88, (1966) 682–721, III, Amer. J. Math. 90, (1969) 55–83, IV, Amer.J.Math. 90, (1969) 1048–1066

[Kd3]    Kodaira, K.: Complex Manifolds and Deformations of Complex Structures, Grundl. Math. Wiss. 283, Springer 1985

[K-O]    Kosarew, S., Okonek, C.: Global moduli spaces and simple holomorphic bundles, Publ. R.I.M.S., Kyoto Univ. 25, (1989) 1–19

[Ku]     Kuranishi, M.: New proof for the existence of locally complete families of complex structures, Proc. Conf. Complex Analysis, Minneapolis, Springer 1965, 142–154

[Kr]     Kurke, H.: Vorlesungen über algebraische Flächen, Teubner-Texte zur Math. 43, Teubner, Leipzig 1982

[LP1]    Le Potier, J.: Fibrés stables de rang 2 sur $\mathbb{P}_2(\mathbb{C})$, Math. Ann. 241, (1979) 217–256

[LP2]    Le Potier, J.: Fibrés vectoriels sur les surfaces K3, Sém. Lelong-Dolbeault-Skoda 1982/1983, Lect. Notes Math. 1028, Springer, Berlin Heidelberg New York 1983

[Li1]    Li, J.: Kodaira dimension of moduli spaces of vector bundles on surfaces, Invent. Math. 115, (1994) 1–40

[Li2]    Li, J.: Algebraic geometric interpretation of Donaldson's polynomial invariants of algebraic surfaces, J. Differ. Geom. 37, (1993) 417–466

[L-Y]   Li, J., Yau, S.T.: Hermitian Yang-Mills connections on non-Kähler manifolds, In: Math. Aspects of String Theory, Ed. S.T. Yau, World Scient. Publ., London 1987, 560–573

[L-Y-Z] Li, J., Yau, S.T., Zheng, F.: A simple proof of Bogomolov's Theorem on class $VII_0$ surfaces with $b_2 = 0$, Illinois J.Math. **34**(2), (1990) 217–220

[Lu1]   Lübke, M.: Chern Klassen von Hermite-Einstein Vektor-Bündeln, Math. Ann. **260**, (1982) 133–141

[Lu2]   Lübke, M.: Stability of Hermitian-Einstein vector bundles, Manuscripta Math. **42**, (1983) 245–257

[L-O1]  Lübke, M., Okonek, C.: Moduli spaces of simple bundles and Hermitian-Einstein connections, Math. Ann. **276**, (1987) 663–674

[L-O2]  Lübke, M., Okonek, C.: Stable bundles on regular elliptic surfaces, J. reine angew. Math. **378**, (1987) 32–45

[ML]    MacLane, S.: Homology, Springer, Berlin Göttingen Heidelberg 1963

[Ma]    Malgrange, B.: Division des distributions, Sém. L.Schwartz 1959/1960, no. 25

[Ml]    Mall, D.: Filtrable bundles of rank 2 on Hopf surfaces, Math. Z. **212**, (1993) 239–256

[Mr1]   Maruyama, M.: Stable vector bundles on an algebraic surface, Nagoya Math. J. **58**, (1975) 25–68

[Mr2]   Maruyama, M.: Openness of a family of torsion-free sheaves, J. Math. Kyoto Univ. **16**, (1976) 627–637

[Mr3]   Maruyama, M.: Moduli of stable sheaves I, J. Math. Kyoto Univ. **17**, (1977) 91–126, II, J. Math. Kyoto Univ. **18**, (1978) 557–614

[Mr4]   Maruyama, M.: On boundedness of families of torsion-free sheaves, J. Math. Kyoto Univ. **21**, (1981) 673–701

[Mr5]   Maruyama, M.: Elementary transformations in the theory of algebraic vector bundles, In: Algebraic Geometry, Eds. J.M. Aroca, R. Buchweitz, M. Giusti, M. Merle, Proc. Conf. La Rábida, Lect. Notes Math. **961**, Springer, Berlin Heidelberg New York 1982, 241–266

[Mr6]   Maruyama, M.: Moduli of stable sheaves-generalities and the curve of jumping lines of vector bundles on $\mathbb{P}^2$, Advanced Studies of Pure Math. I, Alg. Var. and Anal. Var., Kinokuniya and North-Holland 1983, 1–27

[Mr7]   Maruyama, M.: The equations of plane curves and the moduli spaces of vector bundles on $\mathbb{P}^2$, In: Algebraic and topological theories, to the memory of T. Miyata, Tokyo 1985, 430–466

[Mr8]   Maruyama, M.: Vector bundles on $\mathbb{P}^2$ and torsion sheaves on the dual plane, In: Vector Bundles on Algebraic Varieties, Proc. Conf. Bombay 1984, Oxford Univ. Press 1987, 275–339

[Mt]    Matsumura, H.: Commutative Algebra, Benjamin/Cummings Publ. Co. 1980

[Ms]    Matsushima, Y.: Fibrés holomorphes sur un tore complexe, Nagoya Math. J. **14**, (1959) 1–24

[Mj]    Miyajima, K.: Kuranishi family of vector bundles and algebraic description of the moduli space of Einstein-Hermite connections, Publ. R.I.M.S., Kyoto Univ. **25**, (1989) 301–320

[Mi]    Miyaoka, Y.: Kähler metrics on elliptic surfaces, Proc. Japan Acad., Ser.A **50**, (1974) 533–536

[Mu1]   Mukai, S.: Symplectic structure of the moduli space of sheaves on an abelian or K3 surface, Invent. Math. **77**, (1984) 101–116

[Mu2]   Mukai, S.: On the moduli space of bundles on K3 surfaces, In: Vector Bundles on Algebraic Varieties, Proc. Conf. Bombay 1984, Oxford Univ. Press 1987, 341–413

[Mm1]  Mumford, D.: The topology of normal singularities of an algebraic surface and a criterion for simplicity, Publ. Math. I.H.E.S. **9**, (1961)

[Mm2]  Mumford, D.: The canonical ring of an algebraic surface, Ann. Math. **76**, (1962) 612–615

[Mm3]  Mumford, D.: Lectures on curves on an algebraic surface, Ann. Math. Studies **59**, Princeton Univ. Press, Princeton 1966

[Mm4]  Mumford, D.: Abelian varieties, Oxford Univ. Press, 1970

[Na]   Naie, D.: Special rank two vector bundles over Enriques surfaces, Math. Ann. **300**, (1994) 297–316

[Nk]   Nakamura, I.: On surfaces of class $VII_0$ with curves, Invent. Math. **78**, (1984) 393–443

[N-S]  Narasimhan, M.S., Seshadri, C.S.: Stable and unitary vector bundles on a compact Riemann surface, Ann. Math. **82**, (1965) 540–564

[Nw]   Newstead, P.E.: Introduction to Moduli Problems and Orbit Spaces, Tata Inst. Lect. Notes **51**, Springer 1978

[N-R]  Nijenhuis, A., Richardson, R.W.: Cohomology and deformations of algebraic structures, Bull. Amer. Math. Soc. **70**(3), (1964) 406–411

[Nr1]  Norton, A.: Non-separation in the moduli of complex vector bundles, Math. Ann. **235**, (1975) 1–16

[Nr2]  Norton, A.: Analytic moduli of complex vector bundles, Indiana Univ. Math. J. **28**, (1979) 365–387

[O-S-S] Okonek, C., Schneider, M., Spindler, H.: Vector bundles on complex projective spaces, Birkhäuser, Basel Boston Stuttgart 1980

[O-V1] Okonek, C., Van de Ven, A.: Stable bundles and differentiable structures on certain elliptic surfaces, Invent. Math. **86**, (1986) 357–370

[O-V2] Okonek, C., Van de Ven, A.: Vector Bundles and Geometry: some open problems, Math. Göttingen, Heft **17**, (1988)

[O-V3] Okonek, C., Van de Ven, A.: $\Gamma$-type-invariants associated to $PU(2)$-bundles and the differentiable structure of Barlow's surface, Invent. Math. **95**, (1989) 601–614

[O-V4] Okonek, C., Van de Ven, A.: Stable bundles, Instantons and $C^\infty$-structures on algebraic surfaces, In: Encyclopedia of Math. Sci. Vol. **69**,Eds. W. Barth, R. Narasimhan, Several complex variables VI, Springer, Berlin Heidelberg New York London Paris Tokyo Hong Kong Barcelona 1990, 197–249

[Pa]   Pardini, R.: Abelian covers of algebraic varieties, J. reine angew. Math. **417**, (1991) 191–213

[Pe]   Peterson, F.P.: Some remarks on Chern classes, Ann. Math. **69**, (1959) 414–420

[Pl]   Plantiko, R.: Darstellungsräume von Fundamentalgruppen nicht-Kählerscher komplexer Flächen, Ph.D. thesis, Zürich 1994

[Po]   Pourcin, G.: Théorème de Douady au-dessus de S, Ann. Scuola norm. sup. Pisa **23**, (1969) 451–459

[Q1]   Qin, Z.: Birational properties of moduli spaces of stable locally free rank-2 sheaves on algebraic surfaces, Manuscripta Math. **72**, (1991) 163–180

[Q2]   Qin, Z.: Chamber structures of algebraic surfaces with Kodaira dimension zero and moduli spaces of stable rank two bundles, Math. Z. **207**, (1991) 121–136

[Q3]   Qin, Z.: Moduli spaces of stable rank-2 bundles on ruled surfaces, Invent. Math. **110**, (1992) 615–626

[Q4]   Qin, Z.: Equivalence classes of polarizations and moduli spaces of sheaves, J. Differ. Geom. **37**, (1993) 397–415

[Q5]   Qin, Z.: On the existence of stable rank-2 sheaves on algebraic surfaces, J. reine angew. Math. **439**, (1993) 213–219

[Sf1]   Safarevich, I.R., et al.: Algebraic surfaces, Proc. Steklov Inst. Math. **75** (1965),
        A.M.S. Translations Providence, R.I. 1967

[Sf2]   Safarevich, I.R.: Basic Algebraic Geometry, Ed. St. Encicl., Bucharest 1976

[Sh]    Scheja, G.: Fortsetzungssätze der komplex-analytischen Cohomologie und ihre alge-
        braische Charakterisierung, Math. Ann **157**, (1964) 75–94

[Sc]    Schlessinger, M.: Functors of Artin rings, Trans. Amer. Math. Soc. **130**, (1968)
        208–222

[Sn]    Schneider, M.: Chernklassen semi-stabiler Vektorraumbündel vom Rang 3 auf dem
        komplexen projectiven Raum, J. reine angew. Math. **315**, (1980) 211–220

[Sm]    Schumacher, G.: On the geometry of moduli spaces, Manuscripta Math. **50**, (1985)
        229–269

[Ss1]   Schuster, H.W.: Formale Deformationstheorien, Habilitationsschrift, München 1971

[Ss2]   Schuster, H.W.: Locally free resolutions of coherent sheaves on surfaces, J. reine
        angew. Math. **337**, (1982) 159–165

[Sw1]   Schwarzenberger, R.L.E.: Vector bundles on algebraic surfaces, Proc. London Math.
        Soc. **11**(3), (1961) 601–622

[Sw2]   Schwarzenberger, R.L.E.: Vector bundles on the projective plane, Proc. London
        Math. Soc. **11**(3), (1961) 623–640

[Se]    Seiler, W.: Moduli of surfaces of general type with a fibration by genus two curves,
        Math. Ann. **301**, (1995) 771–812

[Sl]    Selder, E.: Picard-Zahlen komplexer Tori, Dissertation, Osnabrück 1981

[So1]   Serrano, F.: Multiple fibres of a morphism, Comment. Math. Helv. **65**, (1990) 287–
        298

[So2]   Serrano, F.: The Picard group of a quasi-bundle, Manuscripta Math. **73**, (1991)
        63–82

[Sr1]   Serre, J.P.: Faisceaux algébriques cohérents, Ann. Math. **61**, (1955) 197–278

[Sr2]   Serre, J.P.: Un théorème de dualité, Comment. Math. Helv. **29**, (1955) 9-26

[Sr3]   Serre, J.P.: Géométrie algébrique et géométrie analytique, Ann. Inst. Fourier **6**,
        (1956) 1–42

[Sr4]   Serre, J.P.: Sur les modules projectifs, Sém. Dubreil-Pisot 1960/1961, Exp. 2, Fac.
        Sci. Paris 1963

[Sr5]   Serre, J.P.: Algèbre locale, multiplicités, Lect. Notes Math. **11**, Springer 1975

[Sd]    Seshadri, C.S.: Theory of Moduli, In: Algebraic Geometry-Arcata 1974, Amer. Math.
        Soc., Proc. Symp. Pure Math. **29**, (1975) 263–304

[Sg]    Siegel, C.L.: Meromorphe Funktionen auf kompakten analytischen Manig-
        faltigkeiten, Nachr. Akad. Wiss. Göttingen, Math. Phys. Kl.II, (1955) 71–77

[Si]    Siu, Y.T.: Every K3 surface is Kähler, Invent. Math. **73**, (1983) 139–150

[S-T]   Siu, Y.T., Trautmann, G.: Gap-sheaves and extensions of coherent analytic sub-
        sheaves, Lect. Notes Math. **172**, Springer 1971

[Sp]    Spanier, E.H.: Algebraic Topology, McGraw-Hill 1966

[St1]   Strømme, S.A.: Deforming vector bundles on the projective plane, Math. Ann. **263**,
        (1983) 385–397

[St2]   Strømme, S.A.: Ample divisors on fine moduli spaces on the projective plane, Math.
        Z. **187**, (1984) 405–423

[Tk]    Takemoto, F.: Stable vector bundles on algebraic surfaces I, Nagoya Math. J. **47**,
        (1972) 29–48, II, Nagoya Math. J. **52**, (1973) 173–195

[Tb1]   Taubes, C.H.: Stability in Yang-Mills Theories, Commun. Math. Phys. **91**, (1983)
        235–263

[Tb2]   Taubes, C.H.: Self-dual connections on 4-manifolds with indefinite intersection ma-
        trix, J. Differ. Geom. **19**, (1984) 517–560

[Te1]   Teleman, A.D.: Stable vector bundles over non-kählerian surfaces, Ph.D. thesis, Zürich 1993

[Te2]   Teleman, A.D.: Projectively flat surfaces and Bogomolov's theorem on class $VII_0$ surfaces, Intern. J. Math. **5**(2), (1994) 253–264

[Tj1]   Tjurin, A.N.: On the classification of two-dimensional fibre bundles over an algebraic curve of arbitrary genus, Izv. Akad. Nauk. SSSR, Ser. Math. **28**, (1964) 21–52

[Tj2]   Tjurin, A.N.: Classification of vector bundles over an algebraic curve of arbitrary genus, Izv. Akad. Nauk. SSSR, Ser. Math. **29**, (1965) 657–688

[To1]   Toma, M.: Une classe de fibrés vectoriels holomorphes sur les 2-tores complexes, C.R. Acad. Sci. Paris, t. **311**, (1990) 257–258

[To2]   Toma, M.: On the existence of simple reducible vector bundles on complex surfaces of algebraic dimension zero, Publ. R.I.M.S. Kyoto Univ. **27**, (1991) 533–550

[To3]   Toma, M.: Holomorphe Vektorbündel auf nichtalgebraischen Flächen, Ph.D. thesis, Bayreuth 1992

[Tt]    Trautmann, G.: Moduli for vector bundles on $\mathbb{P}_n(\mathbb{C})$, Math. Ann. **237**, (1978) 167–186

[Ue1]   Ueno, K.: Classification theory of algebraic varieties and compact complex spaces, Lect. Notes Math. **439**, Springer, Berlin Heidelberg 1975

[Ue2]   Ueno, K.: Kodaira dimension for certain fibre spaces, In: Complex Analysis and Algebraic Geometry, Iwanami-Shoten, Tokyo 1977, 279–292

[U-Y]   Uhlenbeck, K.K., Yau, S.T.: On the existence of Hermitian Yang-Mills connections in stable vector bundles, Commun. Pure Appl. Math. **3**, (1986) 257–293

[Um1]   Umemura, H.: Stable vector bundles with numerically trivial Chern classes over a hyperelliptic surface, Nagoya Math. J. **59**, (1975) 107–134

[Um2]   Umemura, H.: Moduli spaces of the stable vector bundles over abelian surfaces, Nagoya Math. J. **77**, (1980) 47–60

[Vv]    Van de Ven, A.: On the Enriques classification of algebraic surfaces, Sém. Bourbaki, Exp. 506 (1977), Lect. Notes Math. **677**, Springer, Berlin Heidelberg 1978, 237–251

[Vg]    Vogelaar, J.A.: Constructing vector bundles from codimension two subvarieties, Ph.D. thesis, Leiden 1978

[Vu1]   Vuletescu, V.: Sur l'existence des fibrés vectoriels stables de rang deux sur les surfaces non-kähleriennes, C.R. Acad. Sci. Paris, t. **321**, (1995) 591–593

[Vu2]   Vuletescu, V.: An elementary proof of Bănică-Le Potier's inequality $\Delta \geq 0$, Preprint Bucharest 1995

[Vu3]   Vuletescu, V.: Holomorphic vector bundles with zero Chern numbers on nonalgebraic surfaces, Preprint Inst. Mat. Acad. Române **26**, 1995

[Vu4]   Vuletescu, V.: Vectors bundles over surfaces, Ph.D. thesis, Bucharest 1995

[Wa]    Walter, C.H.: Components of the stack of torsion-free sheaves of rank 2 on ruled surfaces, Math. Ann. **301**, (1995) 699–716

[Wh]    Wehler, J.: Moduli space and versal deformation of stable vector bundles, Rev. Roumaine Math. Pures Appl. **30**, (1985) 69–78

[We]    Weil, A.: Variétés Kähleriennes, Act. Sci. Ind. 1267, Hermann, Paris 1958

[Wl]    Wells, R.O.: Differential analysis on complex manifolds, Prentice-Hall, New Jersey 1973

[Wu]    Wu, W.: Sur les espaces fibrés, Act. Sci. Ind. 1183, Hermann, Paris 1952

[Za]    Zariski, O.: Algebraic surfaces, Ergebnisse der Mathematik und ihrer Grenzgebiete **61**, Springer, Berlin Heidelberg 1971

[Zu]    Zuo, K.: Generic smoothness of the moduli spaces of rank two stable vector bundles over algebraic surfaces, Math. Z. **207**, (1991) 629–643

# Index

# Lecture Notes in Mathematics

For information about Vols. 1–1439
please contact your bookseller or Springer-Verlag

Vol. 1479: S. Bloch, I. Dolgachev, W. Fulton (Eds.), Algebraic Geometry. Proceedings, 1989. VII, 300 pages. 1991.

Vol. 1480: F. Dumortier, R. Roussarie, J. Sotomayor, H. Żołądek, Bifurcations of Planar Vector Fields: Nilpotent Singularities and Abelian Integrals. VIII, 226 pages. 1991.

Vol. 1481: D. Ferus, U. Pinkall, U. Simon, B. Wegner (Eds.), Global Differential Geometry and Global Analysis. Proceedings, 1991. VIII, 283 pages. 1991.

Vol. 1482: J. Chabrowski, The Dirichlet Problem with $L^2$-Boundary Data for Elliptic Linear Equations. VI, 173 pages. 1991.

Vol. 1483: E. Reithmeier, Periodic Solutions of Nonlinear Dynamical Systems. VI, 171 pages. 1991.

Vol. 1484: H. Delfs, Homology of Locally Semialgebraic Spaces. IX, 136 pages. 1991.

Vol. 1485: J. Azéma, P. A. Meyer, M. Yor (Eds.), Séminaire de Probabilités XXV. VIII, 440 pages. 1991.

Vol. 1486: L. Arnold, H. Crauel, J.-P. Eckmann (Eds.), Lyapunov Exponents. Proceedings, 1990. VIII, 365 pages. 1991.

Vol. 1487: E. Freitag, Singular Modular Forms and Theta Relations. VI, 172 pages. 1991.

Vol. 1488: A. Carboni, M. C. Pedicchio, G. Rosolini (Eds.), Category Theory. Proceedings, 1990. VII, 494 pages. 1991.

Vol. 1489: A. Mielke, Hamiltonian and Lagrangian Flows on Center Manifolds. X, 140 pages. 1991.

Vol. 1490: K. Metsch, Linear Spaces with Few Lines. XIII, 196 pages. 1991.

Vol. 1491: E. Lluis-Puebla, J.-L. Loday, H. Gillet, C. Soulé, V. Snaith, Higher Algebraic K-Theory: an overview. IX, 164 pages. 1992.

Vol. 1492: K. R. Wicks, Fractals and Hyperspaces. VIII, 168 pages. 1991.

Vol. 1493: E. Benoît (Ed.), Dynamic Bifurcations. Proceedings, Luminy 1990. VII, 219 pages. 1991.

Vol. 1494: M.-T. Cheng, X.-W. Zhou, D.-G. Deng (Eds.), Harmonic Analysis. Proceedings, 1988. IX, 226 pages. 1991.

Vol. 1495: J. M. Bony, G. Grubb, L. Hörmander, H. Komatsu, J. Sjöstrand, Microlocal Analysis and Applications. Montecatini Terme, 1989. Editors: L. Cattabriga, L. Rodino. VII, 349 pages. 1991.

Vol. 1496: C. Foias, B. Francis, J. W. Helton, H. Kwakernaak, J. B. Pearson, $H_\infty$-Control Theory. Como, 1990. Editors: E. Mosca, L. Pandolfi. VII, 336 pages. 1991.

Vol. 1497: G. T. Herman, A. K. Louis, F. Natterer (Eds.), Mathematical Methods in Tomography. Proceedings 1990. X, 268 pages. 1991.

Vol. 1498: R. Lang, Spectral Theory of Random Schrödinger Operators. X, 125 pages. 1991.

Vol. 1499: K. Taira, Boundary Value Problems and Markov Processes. IX, 132 pages. 1991.

Vol. 1500: J.-P. Serre, Lie Algebras and Lie Groups. VII, 168 pages. 1992.

Vol. 1501: A. De Masi, E. Presutti, Mathematical Methods for Hydrodynamic Limits. IX, 196 pages. 1991.

Vol. 1502: C. Simpson, Asymptotic Behavior of Monodromy. V, 139 pages. 1991.

Vol. 1503: S. Shokranian, The Selberg-Arthur Trace Formula (Lectures by J. Arthur). VII, 97 pages. 1991.

Vol. 1504: J. Cheeger, M. Gromov, C. Okonek, P. Pansu, Geometric Topology: Recent Developments. Editors: P. de Bartolomeis, F. Tricerri. VII, 197 pages. 1991.

Vol. 1505: K. Kajitani, T. Nishitani, The Hyperbolic Cauchy Problem. VII, 168 pages. 1991.

Vol. 1506: A. Buium, Differential Algebraic Groups of Finite Dimension. XV, 145 pages. 1992.

Vol. 1507: K. Hulek, T. Peternell, M. Schneider, F.-O. Schreyer (Eds.), Complex Algebraic Varieties. Proceedings, 1990. VII, 179 pages. 1992.

Vol. 1508: M. Vuorinen (Ed.), Quasiconformal Space Mappings. A Collection of Surveys 1960-1990. IX, 148 pages. 1992.

Vol. 1509: J. Aguadé, M. Castellet, F. R. Cohen (Eds.), Algebraic Topology - Homotopy and Group Cohomology. Proceedings, 1990. X, 330 pages. 1992.

Vol. 1510: P. P. Kulish (Ed.), Quantum Groups. Proceedings, 1990. XII, 398 pages. 1992.

Vol. 1511: B. S. Yadav, D. Singh (Eds.), Functional Analysis and Operator Theory. Proceedings, 1990. VIII, 223 pages. 1992.

Vol. 1512: L. M. Adleman, M.-D. A. Huang, Primality Testing and Abelian Varieties Over Finite Fields. VII, 142 pages. 1992.

Vol. 1513: L. S. Block, W. A. Coppel, Dynamics in One Dimension. VIII, 249 pages. 1992.

Vol. 1514: U. Krengel, K. Richter, V. Warstat (Eds.), Ergodic Theory and Related Topics III. Proceedings, 1990. VIII, 236 pages. 1992.

Vol. 1515: E. Ballico, F. Catanese, C. Ciliberto (Eds.), Classification of Irregular Varieties. Proceedings, 1990. VII, 149 pages. 1992.

Vol. 1516: R. A. Lorentz, Multivariate Birkhoff Interpolation. IX, 192 pages. 1992.

Vol. 1517: K. Keimel, W. Roth, Ordered Cones and Approximation. VI, 134 pages. 1992.

Vol. 1518: H. Stichtenoth, M. A. Tsfasman (Eds.), Coding Theory and Algebraic Geometry. Proceedings, 1991. VIII, 223 pages. 1992.

Vol. 1519: M. W. Short, The Primitive Soluble Permutation Groups of Degree less than 256. IX, 145 pages. 1992.

Vol. 1520: Yu. G. Borisovich, Yu. E. Gliklikh (Eds.), Global Analysis – Studies and Applications V. VII, 284 pages. 1992.

Vol. 1521: S. Busenberg, B. Forte, H. K. Kuiken, Mathematical Modelling of Industrial Process. Bari, 1990. Editors: V. Capasso, A. Fasano. VII, 162 pages. 1992.

Vol. 1522: J.-M. Delort, F. B. I. Transformation. VII, 101 pages. 1992.

Vol. 1523: W. Xue, Rings with Morita Duality. X, 168 pages. 1992.

Vol. 1524: M. Coste, L. Mahé, M.-F. Roy (Eds.), Real Algebraic Geometry. Proceedings, 1991. VIII, 418 pages. 1992.

Vol. 1525: C. Casacuberta, M. Castellet (Eds.), Mathematical Research Today and Tomorrow. VII, 112 pages. 1992.